Fourier Transforms in Physics

Student Monographs in Physics

Series Editor: Professor Douglas F Brewer
Professor of Experimental Physics, University of Sussex

Other books in the series:

Microcomputers
 D G C Jones

Maxwell's Equations and their Applications
 E G Thomas and A J Meadows

Oscillations and Waves
 R Buckley

Fourier Transforms in Physics

D C Champeney
School of Mathematics and Physics, University of East Anglia

Adam Hilger Ltd, Bristol and Boston

© D C Champeney 1985

All rights reserved. No part of this publication may be reproduced, stored in a retrieval system or transmitted in any form or by any means, electronic, mechanical, photocopying, recording or otherwise, without prior permission of the publisher.

British Library Cataloguing in Publication Data

Champeney, D. C.
  Fourier transforms in physics.——(Student monographs in physics)
  1. Fourier transformations
  I. Title
  515.7′23 QA403.5

ISBN 0-85274-794-2

Published by Adam Hilger Ltd
Techno House, Redcliffe Way, Bristol BS1 6NX, England
PO Box 230, Accord, MA 02018, USA

Printed in Great Britain by Page Bros (Norwich) Ltd

Contents

Preface vii

1 Introductory Survey
 1.1 Harmonic Oscillations 1
 1.2 Harmonic Synthesis and Analysis 3
 1.3 Physical Applications 9

2 The Fourier Formulae
 2.1 Fourier Series 11
 2.2 Half-range Series 15
 2.3 Complex Notation 16
 2.4 The Fourier Transform 18
 2.5 Parseval's Formulae 22

3 Some Applications
 3.1 Electrical Circuits 24
 3.2 Diffraction 26
 3.3 Oscillations on a String 31
 3.4 Heat Conduction 34
 3.5 The Bandwidth Theorem and Uncertainty Principle 36

4 Further Topics
 4.1 Sampling and Computation 39
 4.2 The Fast Fourier Transform 42
 4.3 Rigorous Fourier Theorems 46

Appendix 1: Some Fourier Series 51
Appendix 2: Some Fourier Transforms 53

Index 55

Preface

This book is intended as an introduction to Fourier series and Fourier transforms for first or second year students on a degree course in physics. The applicability of Fourier techniques is so widespread, however, that students of engineering, applied mathematics, chemistry, and other physical sciences should also find much of the material relevant. No previous knowledge of Fourier methods is assumed, but the reader is expected to be familiar with the standard techniques of integration and to have some understanding of the use of complex numbers. The core material comes in Chapters 2 and 3, which deal respectively with the basic mathematical formulae and with a variety of physical applications, whilst Chapter 1 prepares the way for these. Chapter 4 contains additional material which is presented in rather more condensed form; it deals first with the computation of Fourier transforms, and then concludes with a discussion of the proofs underlying the various Fourier formulae. The section on computation assumes a simple working knowledge of BASIC programming, and is written with a BBC microcomputer in mind, though the results are readily adaptable to other microcomputers. The examples which appear throughout the book are presented in the form of problems and solutions and are intended to complement the brief descriptive sections which precede them.

As with the other books in this series the aim has been to identify an area which students often have difficulty with, and to provide a short and fairly easy route into the subject. Many topics, such as convolution, correlation and noise spectra, lie beyond the scope of this introduction. However, if this book prepares the reader for more advanced treatments and if it gives an appreciation of the widespread applicability of Fourier techniques then it will have served its purpose.

Introductory Survey

1.1 Harmonic Oscillations

Fourier analysis is a mathematical technique that deals with the addition of several sinusoidal or cosinusoidal oscillations to form a resultant. It also deals with the reverse problem of starting from some such resultant and working backwards to find the sine or cosine oscillations from which it was formed. Any branch of physical science in which sinusoidal or cosinusoidal oscillations play a prominent part will probably utilise the theory of Fourier analysis at some stage in its theoretical development. We list some of these topics later in this chapter, but we start by establishing some nomenclature.

If a quantity y varies with time t according to

$$y = C \sin 2\pi f t \tag{1.1}$$

where C and f are constants, we say that y is oscillating sinusoidally with time at *frequency* f and with *amplitude* C. If t is measured in seconds, the unit of f will be the Hertz (Hz). The time T for one complete oscillation is called the *period* of the oscillation, and $T = 1/f$. Sometimes it is convenient to introduce the *angular frequency*, ω, defined as equal to $2\pi f$ and having the units of radians per second. A sinusoidal oscillation is illustrated graphically in figure 1.1(a). Similarly, if

$$y = C \cos 2\pi f t \tag{1.2}$$

we say y is varying cosinusoidally with time at frequency f with amplitude C. Notice from figure 1.1(b) that the sinusoidal oscillation lags behind the cosinusoidal one by a quarter of a period, but otherwise has the same shape.

The oscillation described by

$$y = C \sin(2\pi f t + \phi) \tag{1.3}$$

where ϕ is another constant, is described as a phase shifted sinusoidal oscillation with a phase advance of ϕ radians. As figure 1.1 illustrates, this oscillation is ahead of the pure sinusoidal oscillation by a time $\phi/2\pi f$: we can see this because $(2\pi f t + \phi)$ is zero at a time $t = t_0 = (-\phi/2\pi f)$ rather than when $t = 0$. The oscillation

$$y = C \cos(2\pi f t + \psi) \tag{1.4}$$

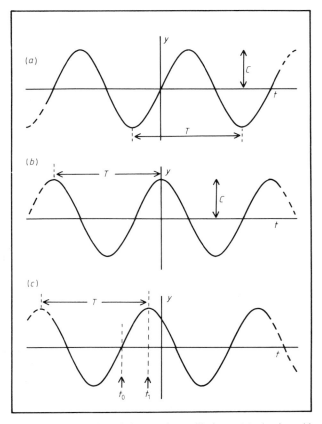

Figure 1.1 Graphs of harmonic oscillations. (*a*) A sinusoid, equation (1.1), (*b*) a cosinusoid, equation (1.2), and (*c*) a harmonic oscillation that can be regarded as a phase shifted sinusoid, equation (1.3), or as a phase shifted cosinusoid, equation (1.4), or as a sum of sinusoid and cosinusoid, equation (1.5). Graph (*c*) is drawn with $\phi = 3\pi/4$ and $\psi = \pi/4$.

where ψ is some constant, is called a phase shifted cosinusoidal oscillation with a phase advance of ψ radians. This is ahead of the pure cosinusoid by a time $\psi/2\pi f$, since $(2\pi f t + \psi)$ is zero when $t = t_1 = (-\psi/2\pi f)$. Clearly, if $\phi = \psi + \pi/2$ then equations (1.3) and (1.4) represent identical oscillations, so we have two ways of describing the same result. A third way of describing this oscillation arises if we use the formula for the sine of a sum of angles as follows:

$$y = C \sin(2\pi f t + \phi)$$

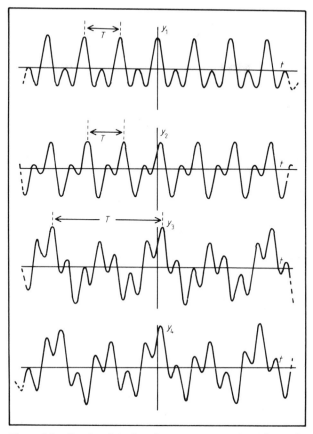

Figure 1.2 Graphs of the functions y_1, y_2, y_3, and y_4 of equation (1.7). y_1 and y_2 have a period $T = 1/3$ s, y_3 has $T = 1$ s, whilst y_4 is not periodic.

We will describe these in turn. The corresponding analysis comes in Chapter 2.

Equation (1.9) implies an infinite set of cosine and sine amplitudes rather than the finite number N in equation (1.8). For instance, an example considered by Fourier is based on the following infinite set of sine coefficients:

$$B_1 = 1, \ B_2 = 1/2, \ B_3 = 1/3, \ldots, B_n = 1/n, \ldots$$

whilst the corresponding cosine coefficients are zero. Equation (1.9) means that the value of y at any fixed value of t is obtained by first finding the 'partial sum', say S_N, of the first N harmonics (as in equation (1.8)), and then defining y as the limiting value of S_N when N tends to infinity. In the example just described, $B_n = 1/n$ and $A_n = 0$, the partial sums for $N = 2$, 5, and 10 are illustrated graphically in figure 1.3. We see that the resultant is approaching closer to the periodically repeated 'sawtooth' oscillation in figure 1.3(d). More examples will appear in Chapter 2.

Equation (1.10) is interpreted as meaning that harmonic oscillations at *all* frequencies are now added together. Instead of the infinite set of *discrete* frequencies $f_1, 2f_1, 3f_1, \ldots$, we now have the 'larger' infinity consisting of a *continuum* of frequencies extending from $f = 0$ to $f = \infty$. The sets of harmonic

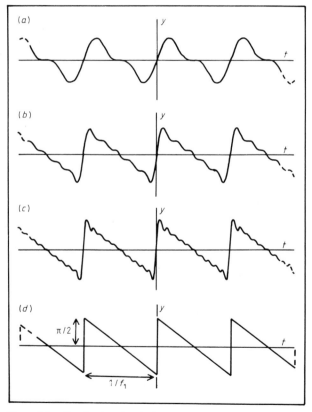

Figure 1.3 Graphs showing the harmonic synthesis of a 'sawtooth'. (a), (b) and (c) represent respectively the sums of the first 2, 5, and 10 terms in equation (1.8), when $A_n = 1/n$ and $B_n = 0$. As more terms are added so the result approaches the sawtooth in (d).

amplitudes A_n and B_n are replaced by functions $A(f)$ and $B(f)$ whose values are defined for every value of f. In any small frequency interval, from $(f - \frac{1}{2}\delta f)$ to $(f + \frac{1}{2}\delta f)$ there are now an infinite number of contributing frequencies, which we assume combine to be equivalent approximately to a harmonic oscillation of frequency f having a cosine amplitude $A(f)\delta f$ and a sine amplitude $B(f)\delta f$. Thus the resultant for the frequency interval of width δf is given approximately as a function of time, for times that are not too large, by:

$$[A(f)\cos(2\pi ft) + B(f)\sin(2\pi ft)]\,\delta f.$$

On adding over all frequency intervals, and taking the limit as $\delta f \to 0$, we obtain the integral in equation (1.10). We say that the resultant, y, has a *continuous amplitude spectrum*, and we refer to the function $A(f)$ as the *cosine amplitude spectrum* of y, whilst $B(f)$ is called the *sine amplitude spectrum* of y.

Example 1.3 *Determine how y varies with time if it has a cosine amplitude spectrum $A(f) = e^{-f}$ ($0 \leq f < \infty$), and if its sine amplitude spectrum is zero.*
Equation (1.10) defines y and we can integrate by parts, remembering that f is the variable of integration with t held constant.

$$y = \int_0^\infty e^{-f} \cos(2\pi ft)\, df$$

$$= [-e^{-f} \cos(2\pi ft)]_0^\infty - \int_0^\infty 2\pi t\, e^{-f} \sin(2\pi ft)\, dt$$

$$= 1 + [2\pi t\, e^{-f} \sin(2\pi ft)]_0^\infty - \int_0^\infty (2\pi t)^2\, e^{-f} \cos 2\pi ft\, df$$

$$= 1 - (2\pi t)^2 y.$$

The answer is thus $y = [1 + (2\pi t)^2]^{-1}$. This is shown graphically in figure 1.4.

Example 1.4 *Determine how y varies with time if it has a cosine amplitude spectrum equal to unity for frequencies up to a cut-off value of 1.0 and equal to zero for higher frequencies, and if the sine amplitude spectrum is zero at all frequencies.*
The integration is easier this time.

$$y = \int_0^\infty A(f) \cos(2\pi ft)\, df$$

$$= \int_0^1 \cos(2\pi ft)\, df$$

$$= \left[\frac{\sin 2\pi ft}{2\pi t}\right]_0^1 = \frac{\sin(2\pi t)}{2\pi t}.$$

This also is shown graphically in figure 1.4.

Notice that in both of these examples y does *not* vary periodically with time. Instead, in each case y dies away to zero as $t \to +\infty$ or $-\infty$. We will call such a quantity a *transient*. This, in fact, is a general principle embodied in the following rule. A quantity varying periodically with time has a discrete amplitude spectrum as in equation (1.9), whilst a transient quantity has a continuous amplitude spectrum as in equation (1.10).

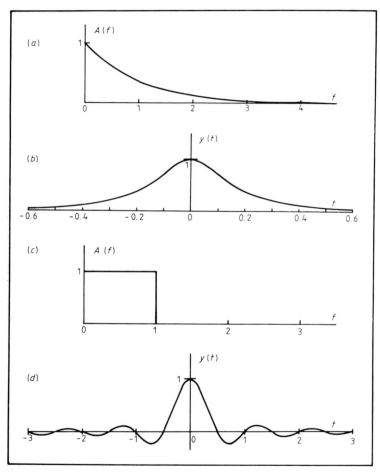

Figure 1.4 Graphs illustrating the harmonic synthesis of transients. The amplitude spectra in (a) and (c) lead respectively to the transients in (b) and (d). The functions shown are:

(a) $A(f) = \exp(-f)$ (b) $y(t) = [1 + (2\pi t)^2]^{-1}$
(c) $A(f) = 1 (f < 1), = 0 (f > 1)$ (d) $y(t) = \sin(2\pi t)/(2\pi t)$.

1.3 Physical Applications

In different fields of application the quantity y, discussed abstractly in §1.2, receives different meanings.

In electrical theory y often represents a voltage or current, and the function $y(t)$ is called a signal. The theory of Fourier analysis is essential to a full understanding of how signals pass through amplifiers and filters and along cables. Oscillatory signals that are periodic have discrete amplitude spectra, whilst a single pulse counts as a transient and has a continuous amplitude spectrum.

In optical theory $y(t)$ represents the oscillatory electric field which exists at any point in space when a light wave passes by. Monochromatic light gives a harmonic variation of y with time, whereas mixtures of colours give a non-harmonic variation. Fourier analysis of $y(t)$ is the mathematical equivalent of breaking light down into its spectrum of colours, achieved physically using a prism. The use of the word spectrum in Fourier theory is borrowed from the language of optics.

In acoustics $y(t)$ will represent the fluctuating pressure which exists at any fixed point in the path of a sound wave. If y varies harmonically with time, then the sound is a 'pure tone', recognised by musicians as a note of definite pitch with no admixture of any other pitch. When the pressure, y, varies in any other way with time, then Fourier analysis of $y(t)$ is the mathematical equivalent of breaking the sound down into a set of pure tones. Harmonic synthesis is the process of combining pure tones of different frequencies to produce a resultant.

Further applications occur in the treatment of other waves including seismic waves, waves on water, vibrations on strings, radio waves and quantum mechanical waves.

Other applications involve replacing the time variable t by a space variable x, and replacing the frequency f by a *spatial frequency s*. The quantity y now has a value dependent on position, and the equations of harmonic synthesis become:

$$y = \sum_{n=1}^{\infty} [A_n \cos(2\pi s_n x) + B_n \sin(2\pi s_n x)]$$

and

$$y = \int_0^{\infty} [A(s) \cos(2\pi s x) + B(s) \sin(2\pi s x)] \, ds.$$

The spatial frequencies s and s_n are equal to $1/\lambda$ or $1/\lambda_n$, where λ and λ_n are the corresponding wavelengths. The units of s are metre^{-1} in the SI system. Sometimes s is called the wavenumber of the harmonic oscillation, whilst $2\pi s$ is called the angular wavenumber with units of radians per unit length.

In the theory of diffraction at an opaque screen with parallel slits, $y(x)$ is used to represent the transmittance of the screen at position x measured across the screen. The transmittance at each point on the screen is the ratio of the transmitted intensity to the incident intensity, and an example appears later in figure 3.1. When $y(x)$ is a periodic function, as with a diffraction grating, the amplitudes A_n and B_n are related to the intensity of the light in the nth order diffraction beam. When there are a finite number of slits then $y(x)$ is a spatial transient and the functions $A(s)$ and $B(s)$ form the starting point of a calculation giving the angular dependence of the intensity of the diffracted light.

In wave mechanics the Schrödinger wave function $\psi(x)$ of a particle moving in one dimension is Fourier analysed in order to find out about the momentum of the particle. The link between spatial frequency $s(=1/\lambda)$ and momentum p is provided by the de Broglie relation $p = h/\lambda = hs$, where h is Planck's constant. If $\psi(x)$ is periodic then the momentum will be one of a discrete set of values, the relative likelihoods being calculable from the amplitudes A_n and B_n. If $\psi(x)$ is a spatial transient, as when the particle is inside a container, then a continuum of momenta is possible and the relative likelihoods of different momenta are calculable from the functions $A(s)$ and $B(s)$. The 'uncertainty principle' of wave mechanics, linking together the uncertainties Δx and Δp in the position and momentum respectively of such a particle, can be precisely expressed using Fourier theory in the form $\Delta x \, \Delta p \geqslant h/4\pi$. The Fourier theory involved is almost identical to that behind the 'bandwidth theorem' which occurs in electrical communication theory (Chapter 3), and here, as in many other instances, two different physical phenomena become related by the Fourier theory used to describe them.

The Fourier Formulae

2.1 Fourier Series

In this Chapter, we describe Fourier's methods for analysing a time dependent quantity into its harmonic components. We deal first with periodically varying quantities, and then in §2.4 we deal with transients.

Here, first, is the famous prescription for analysing a periodic function. We discuss the proof of it in Chapter 4.

Given a quantity $y(t)$ that varies periodically with t with period T, so that $y(t+T) = y(t)$ for any value of t, then provided only that the fluctuations in $y(t)$ are not too pathological it will be possible to write

$$y(t) = \tfrac{1}{2}A_0 + \sum_{n=1}^{\infty} [A_n \cos(2\pi nt/T) + B_n \sin(2\pi nt/T)]. \qquad (2.1)$$

Moreover, the coefficients A_n ($n = 0, 1, 2, \ldots$) and B_n ($n = 1, 2, 3, \ldots$) will be given by

$$A_n = \frac{2}{T} \int_{t_1}^{t_2} y(t) \cos(2\pi nt/T)\, dt \qquad (2.2a)$$

$$B_n = \frac{2}{T} \int_{t_1}^{t_2} y(t) \sin(2\pi nt/T)\, dt \qquad (2.2b)$$

where t_1 and t_2 can be chosen at will, provided $t_2 - t_1 = T$.

The expression on the right of equation (2.1) is called the *Fourier series* of $y(t)$; the quantities A_n and B_n are called the *Fourier series cosine and sine coefficients*, and equations (2.2a,b) are called the *Fourier formulae* for the coefficients. The integrals appearing in the Fourier formulae are examples of *Fourier integrals*. The method can fail if there are values of t around which $y(t)$ either tends to infinity or oscillates with infinite rapidity, and such 'pathological' behaviour is ruled out in the above statement. We discuss this further in §4.3.

The term $\tfrac{1}{2}A_0$ in the Fourier series is included to allow the possibility that $y(t)$ oscillates about some average value other than zero. We write this constant term as $\tfrac{1}{2}A_0$, rather than as A_0, merely because if we do so then the value of A_0 is correctly given by equation (2.2a) along with the cosine coefficients: the cosine of zero is unity, so that on choosing $t_1 = 0$ and $t_2 = T$, the formula gives

$$\tfrac{1}{2}A_0 = \frac{1}{T}\int_0^T y(t)\,\mathrm{d}t.$$

The right-hand side of this equation is precisely how the time average of $y(t)$ is defined and calculated.

Example 2.1 *Find the Fourier series coefficients of the sawtooth of figure 2.1(a) when $h=\pi$ and $T=1$.* In the time interval $t=0$ to 1, y can be expressed by the equation $y=\pi t$; it is thus convenient to choose $t_1=0$ and $t_2=1$. The Fourier formulae, equation (2.2), give

$$A_0 = 2\int_0^1 \pi t\,\mathrm{d}t = \pi.$$

This is as expected, since one can see by inspection that the average value of y is $\pi/2$. For $n \geq 1$ we integrate by parts.

$$A_n = 2\int_0^1 \pi t \cos(2\pi n t)\,\mathrm{d}t$$

$$= [(t/n)\sin(2\pi n t)]_0^1 + \int_0^1 n^{-1}\sin(2\pi n t)\,\mathrm{d}t$$

$$= 0 \qquad \text{for all } n \geq 1.$$

$$B_n = 2\int_0^1 \pi t \sin(2\pi n t)\,\mathrm{d}t$$

$$= [-(t/n)\cos(2\pi n t)]_0^1 + \int_0^1 n^{-1}\cos(2\pi n t)\,\mathrm{d}t$$

$$= -1/n \qquad \text{for all } n \geq 1.$$

The required Fourier series is thus:

$$y(t) = (\pi/2) - \sin(2\pi t) - \frac{\sin(4\pi t)}{2} - \frac{\sin(6\pi t)}{3} - \frac{\sin(8\pi t)}{4} - \cdots.$$

The sum of the first ten terms is shown in figure 2.1(d).

Example 2.2 *Find the Fourier series of the square pulse train of figure 2.1(b) when $h=\pi/2$ and $T=4$.* If we choose $t_1=0$ and $t_2=4$, then the Fourier integrals will each collapse into a sum of two integrals, $\int_0^1$ and $\int_3^4$, because y is zero between $t=1$ and 3. However, if we choose $t_1=-2$ and $t_2=+2$, then the Fourier integrals will collapse into the integral $\int_{-1}^{+1}$ because y is zero between $t=-2$ and -1 and also between $t=+1$ and $+2$. This latter choice of t_1 and t_2 is thus the more convenient, though the results would be the same in both cases. The Fourier formulae, equation (2.2), give

$$A_0 = \tfrac{1}{2} \int_{-1}^{+1} (\pi/2)\, dt = \pi/2$$

as expected for an average value of $\pi/4$. For $n \geq 1$,

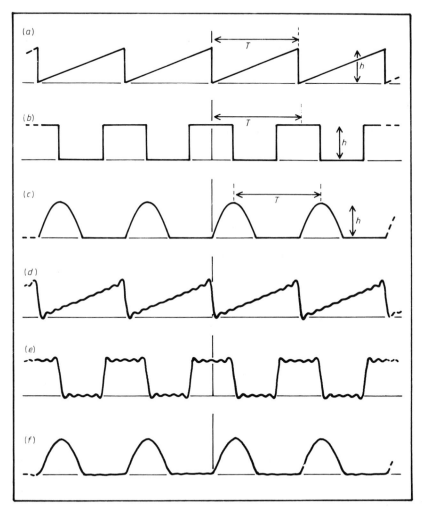

Figure 2.1 Graphs of periodic functions of 'height' h and period T. (a) Sawtooth, (b) square wave, (c) half-wave rectified sinusoid. The sums of the first ten corresponding Fourier harmonics are shown in (d), (e), and (f).

$$A_n = \tfrac{1}{2}\int_{-1}^{+1}(\pi/2)\cos(n\pi t/2)\,dt$$
$$= n^{-1}\sin(n\pi/2)$$

whence $A_1 = 1$, $A_2 = 0$, $A_3 = -1/3$, $A_4 = 0$, $A_5 = +1/5,\ldots$, with all the even coefficients missing (i.e. equal to zero). Similarly

$$B_n = \tfrac{1}{2}\int_{-1}^{+1}(\pi/2)\sin(n\pi t/2)\,dt$$
$$= [(-1/2n)\cos(n\pi t/2)]_{-1}^{+1}$$
$$= 0 \qquad \text{for all } n \geqslant 1.$$

The required Fourier series is thus:

$$y(t) = \pi/4 + \cos(\pi t/2) - \cos(3\pi t/2)/3 + \cos(5\pi t/2)/5 - \cdots$$

the terms being alternately positive and negative. The sum of the first ten terms is shown in figure 2.1(e).

Example 2.3 *Find the Fourier series of the half-wave rectified sine oscillations shown in figure 2.1(c) when $h = \pi/2$ and $T = 2$.* It is equally convenient to choose $t_1 = 0$ with $t_2 = 2$ or $t_1 = -1$ with $t_2 = +1$, because in both cases the Fourier integral collapses to $\int_0^1$, y being zero between $t = -1$ and 0. Inside the interval $t = 0$ to 1 we have $y = (\pi/2)\sin(\pi t)$. The Fourier formulae give

$$A_0 = \int_0^1 (\pi/2)\sin(\pi t)\,dt = 1.$$

For $n \geqslant 1$ integration by parts used twice in succession gives

$$A_n = \int_0^1 (\pi/2)\sin(\pi t)\cos(\pi n t)\,dt$$
$$= \left[\frac{\sin(\pi t)\sin(\pi n t)}{2n}\right]_0^1 - \int_0^1 \frac{\pi\cos(\pi t)\sin(\pi n t)}{2n}\,dt$$
$$= 0 + \left[\frac{\cos(\pi t)\cos(\pi n t)}{2n^2}\right]_0^1 + \int_0^1 \frac{\sin(\pi t)\cos(\pi n t)}{2n^2}\,dt$$
$$= -(1/2n^2)[\cos(\pi n) + 1] + A_n/n^2.$$

Solving for A_n gives, for $n \geqslant 1$,

$$A_n = \frac{-(1 + \cos \pi n)}{2(n^2 - 1)}.$$

$A_1 = 0$, $A_2 = -1/3$, $A_3 = 0$, $A_4 = -1/5$, $A_5 = 0$, $A_6 = -1/35,\ldots$. The sine coefficients are evaluated in a similar fashion:

$$B_1 = \int_0^1 (\pi/2) \sin^2 (\pi t) \, dt$$

$$= (\pi/4) \int_0^1 (1 - \cos 2\pi t) \, dt = \pi/4.$$

For $n \geq 2$,

$$B_n = \int_0^1 (\pi/2) \sin (\pi t) \sin (n\pi t) \, dt$$

$$= (\pi/4) \int_0^1 [\cos (n-1)\pi t - \cos (n+1)\pi t] \, dt$$

$$= \frac{\pi}{4} \left[\frac{\sin (n-1)\pi t}{(n-1)\pi} - \frac{\sin (n+1)\pi t}{(n+1)\pi} \right]_0^1$$

$$= 0 \quad \text{all } n \geq 2.$$

The required Fourier series is thus:

$$y(t) = \frac{1}{2} + \frac{\pi \sin \pi t}{4} - \frac{\cos 2\pi t}{3} - \frac{\cos 4\pi t}{15} - \frac{\cos 6\pi t}{35} - \cdots.$$

The sum of the first ten terms is shown graphically in figure 2.1(f).

The above examples illustrate a symmetry property of Fourier series, a knowledge of which can sometimes save a lot of trouble. If $y(t) = y(-t)$ then the Fourier series will contain only cosine terms; this was the case in example 2.2. In this case $y(t)$ is said to be an *even function* of t, or to have even symmetry. The graph of such a function will superpose on itself when the right-hand portion is 'folded over' on to the left-hand portion. On the other hand, if $y(t) = -y(-t)$, then the function $y(t)$ is said to be an *odd function* of t, or to have odd symmetry; in this case the Fourier series will contain only sine terms. The sawtooth in example 2.1, figure 2.1(a), for instance, consists of the sum of an odd function and a constant term. These results arise because a cosine function has even symmetry whilst a sine function has odd symmetry. The function analysed in example 2.3 was neither odd nor even, and accordingly both sine and cosine terms were necessary.

2.2 Half-range Fourier Series

It is possible to synthesise an arbitrary non-periodic function $y(t)$ using only sine terms, provided that the synthesis is only required to be valid over a limited interval of time, say $t = 0$ to $t = T_1$. The trick is to construct a periodic function of odd symmetry, say $y_{\text{odd}}(t)$, which has period $T = 2T_1$ and which is exactly equal to

$y(t)$ over the limited range $t=0$ to $t=T_1$. Figure 2.2 illustrates this. The Fourier series of y_{odd} is now the required series, containing only sine terms. The series will correctly synthesise $y(t)$ inside the range $t=0$ to T_1, but outside this will yield y_{odd}.

Clearly also, $y(t)$ can be represented by a series containing only cosine terms if we construct an even function $y_{even}(t)$, of period $T=2T_1$, which is equal to $y(t)$ in the range $t=0$ to T_1.

The following equations represent this process, the series being called half-range Fourier cosine and sine series respectively.

$$y = \tfrac{1}{2}A_0 + \sum_{n=1}^{\infty} A_n \cos(\pi n t / T_1) \qquad (0 < t < T_1) \qquad (2.3a)$$

$$= \sum_{n=1}^{\infty} B_n \sin(n\pi t / T_1). \qquad (0 < t < T_1) \qquad (2.3b)$$

$$A_n = (2/T_1) \int_0^{T_1} y(t) \cos(\pi n t / T_1)\, dt \qquad (n = 0, 1, 2, \ldots) \qquad (2.4a)$$

$$B_n = (2/T_1) \int_0^{T_1} y(t) \sin(\pi n t / T_1)\, dt. \qquad (n = 1, 2, 3, \ldots) \qquad (2.4b)$$

2.3 Complex Notation

Complex notation makes the subject seem more difficult in the first instance, but the elegance and economy of space which it brings is ultimately an advantage. With $i = \sqrt{-1} = -1/i$ we can translate sine and cosine functions into complex notation using

$$\cos\theta = (1/2)(e^{i\theta} + e^{-i\theta})$$
$$\sin\theta = (-i/2)(e^{i\theta} - e^{-i\theta}). \qquad (2.5)$$

Likewise, we can translate back from complex exponentials to sines and cosines using

$$\exp(i\theta) \equiv e^{i\theta} = \cos\theta + i\sin\theta$$
$$\exp(-i\theta) \equiv e^{-i\theta} = \cos\theta - i\sin\theta.$$

We thus arrive at yet another way of expressing a harmonic oscillation in addition to equations (1.3)–(1.5), namely

$$y = A\cos(2\pi f t) + B\sin(2\pi f t) \qquad (1.5)$$

$$= C^+ \exp(i 2\pi f t) + C^- \exp(-i 2\pi f t) \qquad (2.6)$$

where

$$C^+ = \tfrac{1}{2}(A - iB) \qquad C^- = \tfrac{1}{2}(A + iB). \qquad (2.7)$$

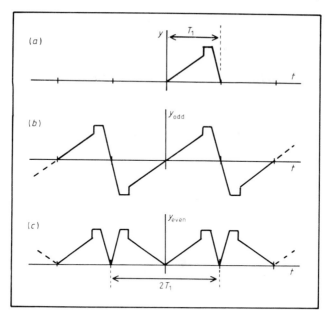

Figure 2.2 Half-range Fourier series. (a) An arbitrary function $y(t)$ defined only between $t = 0$ and $t = T_1$, with (b) an odd periodic function of period $2T_1$, and (c) an even periodic function of period $2T_1$, all of which are equal in value between $t = 0$ and $t = T_1$.

In place of the two real amplitudes A and B we now have the two complex quantities C^+ and C^-, and equation (2.6) is our new representation of a harmonic oscillation. The reader will readily verify (2.6) on substituting (2.5) into (1.5) and comparing with (2.6). We call C^+ the amplitude at frequency $+f$ in the complex exponential representation, and we call C^- the amplitude at frequency $-f$ in the complex exponential representation. Note two things. First, when using ordinary sine and cosine representations we use only positive frequencies; the use of negative frequencies is merely a convenient way of describing the two parts of equation (2.6) based on complex exponentials. Second, equation (2.6) is to be used as an equality: if y is a real quantity we insist that the left and right sides of the equation are *equal*, and we do *not* mean that y is merely the real part of the right-hand side. When y is real it is necessary that the complex numbers C^+ and C^- combine with the complex parts of the exponentials in such a way that the resultant is real.

The Fourier series, equation (2.1), can easily be translated into complex notation to yield

$$y(t) = \tfrac{1}{2}A_0 + \sum_{n=1}^{\infty} \left[C_n^+ \exp(\mathrm{i}2\pi n t / T) + C_n^- \exp(-\mathrm{i}2\pi n t / T) \right]$$

where $C_n^\pm = \frac{1}{2}(A_n \mp iB_n)$. However, it is convenient to rewrite this as a summation over positive, negative and zero values of the integer n:

$$y(t) = \sum_{n=-\infty}^{+\infty} Y_n \exp(+i2\pi nt/T) \qquad (2.8)$$

where

$$\begin{aligned} Y_n &= \tfrac{1}{2}(A_n - iB_n) & n &\geq 1 \\ Y_0 &= \tfrac{1}{2}A_0 & & \\ Y_{-n} &= \tfrac{1}{2}(A_n + iB_n) & n &\geq 1. \end{aligned} \qquad (2.9)$$

For future reference, we solve these to give

$$\begin{aligned} A_n &= Y_n + Y_{-n} & n &\geq 1 \\ B_n &= i(Y_n - Y_{-n}) & n &\geq 1 \\ A_0 &= 2Y_0. & & \end{aligned} \qquad (2.10)$$

Another convenience of the complex notation is that the Fourier formulae, equation (2.2), translate using (2.9) into the single formula:

$$Y_n = \frac{1}{T} \int_{t_1}^{t_2} y(t) \exp(-i2\pi nt/T) \, dt. \qquad (2.11)$$

Example 2.4 *Obtain the Fourier series coefficients, Y_n, of the sawtooth in example 2.1 using complex exponentials.* Using equation (2.11), for $n \neq 0$,

$$\begin{aligned} Y_n &= \int_0^1 \pi t \exp(-i2\pi nt) \, dt \\ &= \left[\frac{-t \exp(-i2\pi nt)}{2in} \right]_0^1 - \int_0^1 \frac{\exp(-i2\pi nt)}{2in} \, dt \\ &= \frac{1}{-2in} + \left[\frac{\exp(-2\pi nt)}{-4\pi n^2} \right]_0^1 \\ &= +i/2n \qquad n = \pm 1, \pm 2, \pm 3, \ldots. \end{aligned}$$

$$Y_0 = \int_0^1 \pi t \, dt = \pi/2.$$

These values are consistent with the previously obtained values of A_n and B_n.

2.4 The Fourier Transform

We now give Fourier's method for analysing a transient into a summation over a continuum of frequencies. We discuss the proof in chapter 4. *Given a quantity $y(t)$*

which decays sufficiently rapidly as $t \to +\infty$ and $t \to -\infty$, and supposing the fluctuations are not too pathological, we can write

$$y(t) = \int_0^\infty [A(f) \cos 2\pi ft + B(f) \sin 2\pi ft] \, df. \tag{2.12}$$

Moreover, the amplitudes $A(f)$ and $B(f)$ will be given by

$$A(f) = 2 \int_{-\infty}^\infty y(t) \cos(2\pi ft) \, dt \qquad (0 \leqslant f < \infty) \tag{2.13}$$

$$B(f) = 2 \int_{-\infty}^\infty y(t) \sin(2\pi ft) \, dt \qquad (0 < f < \infty). \tag{2.14}$$

In the analysis of transients the use of complex exponential notation is very common, and the role played by the amplitudes $A(f)$ and $B(f)$ in the above formulation is now played by a single function $Y(f)$. We find in this alternative formulation that the transient $y(t)$ can be expressed as

$$y(t) = \int_{-\infty}^{+\infty} Y(f) \exp(i2\pi ft) \, df \tag{2.15}$$

where, for positive and negative f, $Y(f)$ is given by

$$Y(f) = \int_{-\infty}^{+\infty} y(t) \exp(-i2\pi ft) \, dt. \tag{2.16}$$

The relation between the amplitude $Y(f)$ and the amplitudes $A(f)$ and $B(f)$ is as follows, and may be obtained readily using the ideas in §2.3.

$$\left. \begin{array}{l} Y(f) = \tfrac{1}{2}[A(f) - iB(f)] \qquad f > 0 \\ Y(0) = \tfrac{1}{2}A(0) \\ Y(-f) = \tfrac{1}{2}[A(f) + iB(f)] \qquad f > 0. \end{array} \right\} \tag{2.17}$$

$$\left. \begin{array}{l} A(f) = Y(f) + Y(-f) \qquad f \geqslant 0 \\ B(f) = i[Y(f) - Y(-f)] \qquad f > 0. \end{array} \right\} \tag{2.18}$$

The quantity $Y(f)$, defined for negative as well as positive f by equation (2.16), is called the *Fourier transform of* $y(t)$. The integrals in equations (2.15), (2.16) are examples of *Fourier integrals*. The first equation represents synthesis of $y(t)$ from complex exponential harmonics, the second equation, (2.16), represents analysis of $y(t)$.

Example 2.5 *A transient quantity y is equal to* e^{-t} *for* $t \geqslant 0$ *and is zero for* $t < 0$; *express y as a sum of complex exponentials and also as a sum of sine and cosine oscillations.* Since y is zero for $t < 0$ the Fourier integral $\int_{-\infty}^{+\infty}$ collapses to $\int_0^\infty$ as follows.

$$Y(f) = \int_{-\infty}^{+\infty} y(t) \exp(-i2\pi ft) \, dt$$

$$= \int_{0}^{\infty} \exp(-i2\pi ft - t) \, dt$$

$$= \left[\frac{\exp(-2\pi ift - t)}{-(2\pi if + 1)} \right]_{0}^{\infty}$$

$$= \frac{1}{1 + i2\pi f} = \frac{1 - i2\pi f}{1 + (2\pi f)^2} \qquad -\infty < f < +\infty \tag{2.19}$$

Using equation (2.18), with $f \geqslant 0$,

$$A(f) = Y(f) + Y(-f) = 2/[1 + (2\pi f)^2] \tag{2.20}$$

$$B(f) = i[Y(f) - Y(-f)] = 4\pi f/[1 + (2\pi f)^2]. \tag{2.21}$$

In figure 2.3(a)–(c) we give graphs of $y(t)$, $A(f)$, $B(f)$ and of the real and imaginary parts of $Y(f)$. Note that $A(f)$ and $B(f)$ are real, even though $Y(f)$ is complex.

Example 2.6 *A transient quantity $y(t)$ is equal to $\tfrac{1}{2}$ for $-1 < t < 1$, and is zero otherwise; find the Fourier transform of y and the cosine and sine amplitude spectra.* The Fourier integral $\int_{-\infty}^{+\infty}$ becomes $\int_{-1}^{+1}$, because y is zero when $|t| > 1$. Thus, using (2.16),

$$Y(f) = \int_{-1}^{+1} (1/2) \exp(-i2\pi ft) \, dt$$

$$= \left[\frac{\exp(-i2\pi ft)}{-i4\pi f} \right]_{-1}^{+1}$$

$$= \frac{i}{2} \left(\frac{\exp(-i2\pi ft) - \exp(i2\pi ft)}{2\pi f} \right)$$

$$= (\sin 2\pi f)/2\pi f \qquad -\infty < f < +\infty. \tag{2.22}$$

Whence from equation (2.18), for $f \geqslant 0$,

$$A(f) = Y(f) + Y(-f) = (\sin 2\pi f)/\pi f \tag{2.23}$$

$$B(f) = i[Y(f) - Y(-f)] = 0. \tag{2.24}$$

In figure 2.3(d)–(f) we give graphs of $y(t)$, $A(f)$ and $Y(f)$.

The above two examples illustrate a feature which is always true when $y(t)$ is itself a real quantity: if $y(t)$ is an even function of time so that $y(t) = y(-t)$ then $Y(f)$ will be real for all f, and also the sine amplitude will be zero. Likewise, if $y(t)$ is a real and odd function of time, i.e. $y(t) = -y(-t)$, then $Y(f)$ will be pure imaginary at all frequencies and the cosine amplitude $A(f)$ will be zero. If $y(t)$ is

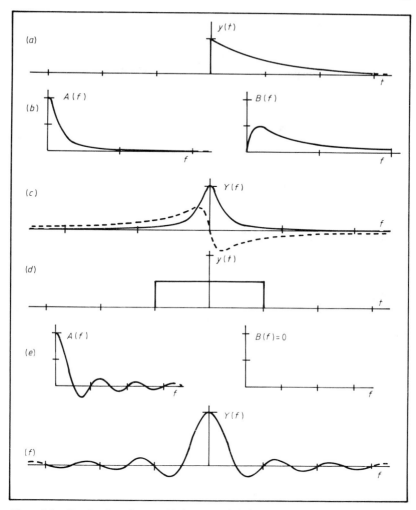

Figure 2.3 Graphs of (a) the one-sided exponential of example 2.5, with (b) its cosine and sine amplitude spectra, equations (2.20), (2.21), and (c) its Fourier transform, equation (2.19). Also (d) the rectangular pulse of example 2.6 with (e) its corresponding spectra, equations (2.22)–(2.24), and (f) its Fourier transform. The complex part of $Y(f)$ is shown by the dashed line in (c). The axes are marked in units in each case.

neither odd nor even, as in example 2.5, then $Y(f)$ will contain a real and an imaginary part, and both $A(f)$ and $B(f)$ will exist.

2.5 Parseval's Formulae

In practical applications involving a periodic quantity $y(t)$ of period T, there is often a need to know the value of the mean squared value, $\overline{y^2}$, defined by

$$\overline{y^2} = \frac{1}{T}\int_0^T y^2(t)\,dt. \tag{2.25}$$

For instance, if $y(t)$ represents the voltage at time t across a resistance R, then the instantaneous power dissipation is y^2/R and the time averaged power is $\overline{y^2}/R$. Correspondingly, if $y(t)$ is a transient, then there is often a need to know the value, say K, of the integral

$$K = \int_{-\infty}^{+\infty} y^2(t)\,dt. \tag{2.26}$$

When y represents the transient voltage across a resistance R, then the total energy dissipated over all time will be K/R.

Parseval's two formulae show how the quantities in equations (2.25) and (2.26) may be evaluated from the Fourier series coefficients, or the Fourier transform, of $y(t)$ as appropriate.

Let us note first that if $y(t)$ varies harmonically with t, with cosine and sine amplitudes A and B (equation (1.5)), then the integral in equation (2.25) can be evaluated directly to give a value $\frac{1}{2}(A^2 + B^2)$ for $\overline{y^2}$. It is then plausible to suggest that if $y(t)$ consists of a Fourier series whose harmonics have cosine and sine amplitudes A_n and B_n, equation (2.1), then the contributions will be additive, so that in total

$$\overline{y^2} = (A_0^2/4) + \sum_{n=1}^{\infty}(A_n^2 + B_n^2)/2. \tag{2.27}$$

The $A_0^2/4$ term arises because of the constant term $A_0/2$ in the Fourier series. This equation is known as the Parseval formula for Fourier series, after the mathematician who established rigorously a set of conditions for its applicability. The formula only fails when the graph of $y(t)$ has pathological fluctuations.

If complex Fourier coefficients Y_n are used, then using equation (2.9) we may readily adapt the above formula to the neat form

$$\overline{y^2} = \sum_{n=-\infty}^{+\infty} |Y_n|^2. \tag{2.28}$$

If the quantity $y(t)$ is itself complex, then we must modify the above Parseval formulae by replacing $y(t)$, A_0, A_n, and B_n by their moduli $|y(t)|$, $|A_0|$, $|A_n|$, and $|B_n|$ whenever they occur.

The physical interpretation of Parseval's Fourier series formula is that if the power dissipation is proportional to $y^2(t)$, then we may associate with the nth harmonic a fraction $(A_n^2 + B_n^2)/2\overline{y^2}$ of the total power. This fraction may alternatively be written as $(|Y_n|^2 + |Y_{-n}|^2)/\overline{y^2}$.

There is a corresponding Parseval formula for a transient $y(t)$ having a Fourier transform with cosine and sine amplitudes $A(f)$ and $B(f)$, or a complex exponential amplitude $Y(f)$, as in §2.4. The result is

$$\int_{-\infty}^{+\infty} y^2(t)\,dt = \int_0^{+\infty} \tfrac{1}{2}[A^2(f) + B^2(f)]\,df$$

$$= \int_{-\infty}^{+\infty} |Y(f)|^2\,df. \qquad (2.29)$$

This formula only fails if the graph of $y(t)$ decays too slowly as $t \to \pm\infty$, and/or it displays pathological fluctuations. If $y(t)$ is itself complex then $y(t)$, $A(f)$, and $B(f)$ must be replaced by their moduli. The interpretation in physical applications is now as follows. If the rate of energy transfer associated with some transient quantity $y(t)$ is proportional to $|y(t)|^2$, so that the total energy is proportional to K, equation (2.26), then we may associate with the narrow frequency range f to $f + \delta f$ an amount of energy that is the following fraction of the total energy:

$$(2K)^{-1}(|A(f)|^2 + |B(f)|^2)\,\delta f = K^{-1}(|Y(f)|^2 + |Y(-f)|^2)\,\delta f.$$

Example 2.7 *Verify the validity of Parseval's formula for the transient* $y(t) = e^{-t}$ $(t \geq 0)$, $y(t) = 0$ $(t < 0)$. The Fourier transform is given in equation (2.19); we may thus evaluate each side of equation (2.29) as follows.

$$\int_{-\infty}^{+\infty} y^2(t)\,dt = \int_0^{+\infty} \exp(-2t)\,dt = \tfrac{1}{2}.$$

$$\int_{-\infty}^{+\infty} |Y(f)|^2\,dt = \int_{-\infty}^{+\infty} |1 + i2\pi f|^{-2}\,df$$

$$= \int_{-\infty}^{+\infty} (1 + 4\pi^2 f^2)^{-1}\,df$$

$$= (2\pi)^{-1} \int_{-\infty}^{+\infty} (1 + \omega^2)^{-1}\,d\omega$$

$$= (2\pi)^{-1}[\tan^{-1}\omega]_{-\infty}^{+\infty} = \tfrac{1}{2}.$$

This is the required result.

Some Applications 3

3.1 Electrical Circuits

Standard AC (alternating current) theory, with which we assume some familiarity, deals with the relations between currents and voltages in a network of components when the currents and voltages vary harmonically with time. We now show how Fourier theory can be used to extend this treatment to deal with waveforms that are not harmonic, such as a transient or, say, a periodic sawtooth.

AC theory relies on the assumption that if a harmonic current, say $I_0 \cos(2\pi ft)$, flows through a component or circuit, then the associated voltage will also be harmonic, say $V_0 \cos(2\pi ft + \alpha)$, where the ratio V_0/I_0 and the phase advance α each depend only on the frequency for a given circuit. Circuits composed of purely resistive, capacitative or inductive components satisfy this assumption, whilst circuits containing diodes, for instance, do not. However, if a non-harmonic current waveform is passed through a circuit satisfying the above assumptions the associated voltage waveform will be distorted in shape, and cannot be calculated simply by scaling and time shifting the current waveform. AC theory seems to be inapplicable.

However, AC theory can be used after all if the non-harmonic current is Fourier analysed into its harmonic components. AC theory is then applied to each component to obtain the corresponding harmonic voltage component and these are then Fourier synthesised to give the actual voltage. The complex impedance Z of a circuit is particularly convenient when used with the complex exponential Fourier integrals, and we now show why.

For given values of phase advance α, defined above, and of the ratio $r = V_0/I_0$, a complex current $I \exp(j\omega t)$ will give rise to a complex voltage $rI \exp(j\omega t \pm j\alpha)$, where in conformity with electrical engineering practice we use j (instead of i) to represent $\sqrt{-1}$: this leaves the symbol i available to represent current. Note that if α is to represent a phase *advance* it is necessary to use the positive sign, $+j\alpha$, when $\omega > 0$, and to use $-j\alpha$ when $\omega < 0$. We now define the complex impedance at any frequency as $Z = r \exp(\pm j\alpha)$, where the positive sign is used for $\omega > 0$ and the negative for $\omega < 0$. With this definition the complex voltage is equal to the

Some Applications

product of the complex current and Z. The well known values of $j\omega L$ and $(-j/\omega C)$ for the impedances in ohms respectively of an inductance L henrys or a capacitor C farads at angular frequency ω rad s^{-1} arise from this definition of impedance, and are valid for negative as well as positive values of ω.

When dealing with filters, a complex quantity $H(\omega)$, called the *transfer function*, plays a role analogous to that of impedance. If a harmonic input signal (current or voltage) leads to a harmonic output signal (current or voltage), then $H(\omega)$ is defined as $r \exp(\pm j\alpha)$ where α is the phase advance of the output relative to the input at angular frequency ω, r is the value of output amplitude divided by input amplitude; the positive and negative signs refer to positive and negative values of ω respectively, as usual.

Example 3.1 *A voltage waveform consisting of a sawtooth, figure 2.1(a), of height $h = 1$ V and period $T = 1$ ms is applied to an idealised low-pass filter which passes frequencies less than 3.5 kHz without change in amplitude or phase, but which removes higher frequencies. Determine the output voltage waveform.* From example 2.1 the cosine and sine Fourier amplitudes of the input waveform (in volts) are $A_0 = 1$, $B_n = -(\pi n)^{-1}$ ($n = 1, 2, 3, \ldots$), $A_n = 0$ ($n \geq 1$). The frequency of the nth harmonic is $n/T = n$ kHz, so the output voltage $v(t)$ only contains harmonic components up to and including $n = 3$, since the filter removes higher frequencies. Thus, as our answer,

$$v(t) = 1/2 - \sin(2\pi t/T)/\pi - \sin(4\pi t/T)/2\pi - \sin(6\pi t/T)/3\pi.$$

Example 3.2 *A voltage square wave as in figure 2.1(b), oscillating between zero and h V and of repetition frequency f_1, is applied across a series combination of capacitance C and inductance L. If $h = 1$ V, $f_1 = 1$ kHz, $4\pi^2 f_1^2 LC = 1/8.9$, and $2\pi f_1 L = 20$ Ω, show that the current will be approximately harmonic and determine its frequency, amplitude and phase.* Let us use the complex exponential Fourier series throughout, V_n and I_n being the nth Fourier coefficients of the voltage and current respectively, for $n = 0, \pm 1, \pm 2, \ldots$. If Z_n is the complex impedance of the LC combination at a frequency equal to the nth harmonic of the square wave, then $I_n = V_n/Z_n$. Values for V_n come from example 2.2 together with equation (2.9), whilst the Z_n come from the general expression $Z = j[\omega L - (\omega C)^{-1}]$ at angular frequency $\omega = n2\pi f_1$. We have $V_0 = h/2$ and $|Z_0| = \infty$ so $I_0 = 0$. For $n > 0$ and $n < 0$ we have

$$V_n = (h/\pi n)\sin(n\pi/2)$$

$$Z_n = j[2\pi n f_1 L - (2\pi n f_1 C)^{-1}]$$

$$I_n = \frac{-j(h/\pi)\sin(n\pi/2)}{2\pi f_1 L[n^2 - (4\pi^2 f_1^2 LC)^{-1}]}$$

$$= \frac{(-j/\pi)\sin(n\pi/2)}{20(n^2 - 8.9)} \text{ amperes}.$$

The $\sin(n\pi/2)$ term shows that I_n is zero unless n is odd, and examination of the expression $(n^2 - 8.9)$ for various values of n shows that the terms I_3 and I_{-3} $(= -I_3)$ dominate the others. In the Fourier series expansion of the current $i(t)$ we can, as an approximation, ignore all terms other than those for $n = \pm 3$, obtaining

$$i(t) = I_3 \exp(6\pi j f_1 t) + I_{-3} \exp(-6\pi j f_1 t)$$
$$= 2jI_3 \sin(6\pi f_1 t)$$
$$= -\pi^{-1} \sin(6\pi f_1 t) \text{ amperes}.$$

Thus the current is approximately harmonic, at a frequency $3f_1 = 3$ kHz, with an amplitude π^{-1} A, and a phase such that the current is zero and decreasing at the mid point of a voltage pulse.

Example 3.3 *A band-pass filter will pass signals in a frequency band of width Δ centred on frequency f_0, but will not pass signals at other frequencies. A steady voltage h is applied to the input between time $-T$ and time $+T$, the input voltage being zero at other times. If the output voltage is $v_2(t)$ and if $\Delta \ll 1/T$, show that as T is varied so $\int_{-\infty}^{+\infty} [v_2(t)]^2 \, dt$ will be proportional to $h^2 \sin^2(2\pi f_0 T)$, the constant of proportionality depending only on the filter characteristic.* The input voltage is a transient, so we use Fourier transforms rather than series. Let the input and output voltages, $v_1(t)$ and $v_2(t)$, have Fourier transforms $V_1(f)$ and $V_2(f)$. If the transfer function of the filter is $H(f)$ then $V_2(f) = V_1(f) H(f)$. From example 2.6 we have

$$V_1(f) = (h/\pi f) \sin(2\pi f T)$$

whose graph is similar to that of figure 2.3(f). Now the condition $\Delta \ll 1/T$ means that at fixed T the quantity $\sin(2\pi f T)$ will change only slightly when f changes by Δ. Moreover, since $H(f)$ is zero except within bands of width Δ at $\pm f_0$, this means we can approximate $V_2(f)$ as follows:

$$V_2(f) = (h/\pi f_0) \sin(2\pi f_0 T) H(f).$$

From Parseval's theorem we now have

$$\int_{-\infty}^{+\infty} |v_2(t)|^2 \, dt = \int_{-\infty}^{+\infty} |V_2(f)|^2 \, df$$
$$= (h/\pi f_0)^2 \sin^2(2\pi f_0 T) \int_{-\infty}^{+\infty} |H(f)|^2 \, df.$$

This establishes the result. Note that the output in this approximation is zero when $2T$ is a multiple of $1/f_0$.

3.2 Diffraction

We now show how Fourier transforms are useful in describing the diffraction of

Some Applications

light when it passes through narrow slits. The treatments of acoustic, x-ray and microwave diffraction, and indeed of all forms of wave diffraction, are very similar. We assume the reader is familiar with the way in which Huygens' principle can be used to describe the diffraction at a pair of narrow parallel slits (Young's experiment), and we extend the description to deal with several parallel slits which are neither narrow, equally spaced, nor equal in width.

Suppose, as an example, we consider three long, parallel slits in an otherwise opaque screen, the widths of the slits being W_1, W_2, and W_3 and the widths of the opaque regions between slits being d_1 and d_2. This is shown in figure 3.1(a), each of the slits extending in and out of the paper. We suppose also that monochromatic light of wavelength λ is incident normally on the screen from the left, and we wish to calculate how the diffracted radiation spreads out on the right-hand side. Choosing an arbitrary reference point O in the plane of the screen, somewhere within the slit system, let $I(\theta)$ be the light intensity at a point P which is at a distance R from O in a direction making an angle θ with the forward normal as shown. If R is large compared with the total width of the slit system then $I(\theta)$ will depend only very slightly on the choice of origin O, and the variation of $I(\theta)$ with θ, at fixed R, will describe the angular distribution of diffracted intensity. We wish to relate $I(\theta)$ to W_1, W_2, W_3, d_1 and d_2.

Our starting point is a function $t(x)$ called the *transmission function* of the diffracting screen; its graph is shown in figure 3.1(b) for the present example. The variable x measures distance across the slit system from the origin O, and at each value of x the value of $t(x)$ is either unity (when x describes a point lying within a slit) or zero (when x describes a point on an opaque portion of the screen). The function $t(x)$ thus summarises succinctly all the information we need about the widths and spacings of the slits. The transmission function is a spatial transient and can thus be Fourier synthesised from complex exponentials covering a continuum of spatial frequencies s:

$$t(x) = \int_{-\infty}^{+\infty} T(s) \exp(2\pi i s x) \, ds. \tag{3.1}$$

The variable s has the units of inverse length, and $T(s)$ is the harmonic amplitude corresponding to s. $T(s)$ can be calculated from $t(x)$ using Fourier's formula,

$$T(s) = \int_{-\infty}^{+\infty} t(x) \exp(-2\pi i s x) \, dx. \tag{3.2}$$

However, both s and $T(s)$ can be given a different interpretation. If, for any diffraction angle θ, we set $s = \lambda^{-1} \sin \theta$ and evaluate $|T(s)|^2$, then we find that as θ is varied, so

$$I(\theta) \propto |T(s)|^2 \tag{3.3}$$

where the constant of proportionality depends only on λ, R and the incident

28 *Fourier Transforms in Physics*

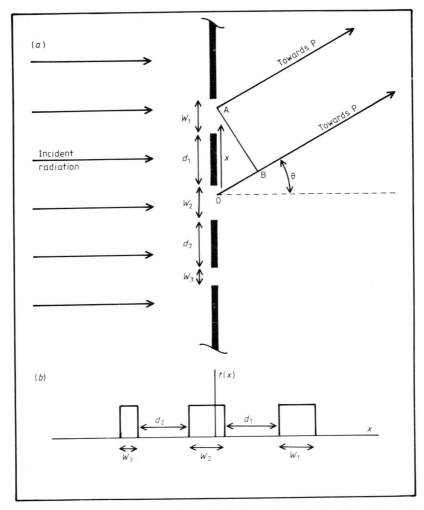

Figure 3.1 (a) Diagram for calculating the diffraction at three slits of widths W_1, W_2 and W_3, each slit running perpendicular to the plane of the diagram. (b) The graph of the corresponding transmission function $t(x)$.

intensity. This then is the answer to our problem: to determine the angular distribution of diffracted intensity we simply 'Fourier transform the transmission function of the screen, and then square the modulus of this transform'. We derive this result in a moment, but first give some examples.

Some Applications

Example 3.4 *Determine the angular distribution of diffracted intensity when green light of wavelength $\lambda = 0.55$ μm is diffracted at a single slit of width $W = 0.5$ mm.* Choose the origin O at the centre of the slit so that the transmission function consists of a rectangular 'pulse' with edges at $x = \pm W/2$. The Fourier transform of such a pulse has been evaluated in example 2.6, and in our present notation

$$|T(s)|^2 = \left(\frac{\sin(\pi s W)}{\pi s}\right)^2.$$

The graph of $|T(s)|^2$ is shown in figure 3.2(a), and we see that the diffraction pattern consists of a broad forward peak around $s = 0$ (i.e. $\theta = 0$), with subsidiary maxima which get weaker at larger s (i.e. larger θ). The zeros in intensity occur at angles such that $\pi s W$ is a multiple of π, that is when $W \sin\theta = n\lambda$ ($n = \pm 1, \pm 2, \ldots$). For $\lambda = 0.55$ μm and $W = 0.5$ mm the first minimum occurs at an angle of 0.063 degrees.

Example 3.5 *Determine the angular distribution of diffracted light when green light, $\lambda = 0.55$ μm, is diffracted at two parallel slits each of width $W = 0.1$ mm, the centres of the slits being a distance $D = 0.5$ mm apart.* If we choose the origin to lie midway between the slits, then the transmission function consists of rectangular pulses centred on $x = \pm D/2$, and $t(x)$ is unity for $(D-W)/2 < |x| < (D+W)/2$ and is zero otherwise. Making use of the fact that $t(x)$ is an even function of x, equation (3.2) becomes

$$\begin{aligned} T(s) &= 2\int_{(D-W)/2}^{(D+W)/2} \cos(2\pi s x)\,dx \\ &= \frac{\sin \pi s(D+W)}{\pi s} - \frac{\sin \pi s(D-W)}{\pi s} \\ &= 2(\pi s)^{-1} \sin(\pi W s)\cos(\pi D s). \end{aligned}$$

The graph of $|T(s)|^2$ is shown in figure 3.2(b) for $D = 5W$. The closely spaced maxima occur at angles such that $\pi D s$ is a multiple of π, that is when $D\sin\theta = n\lambda$, this being the well known formula for Young's double slit experiment. However, the maxima become gradually less intense with increasing angle and the intensity is zero when $\pi W s$ is a multiple of π, that is when $W\sin\theta = m\lambda$ ($m = \pm 1, \pm 2, \ldots$). With the numbers given, the first maximum is at an angle 0.063 degrees, and the fifth such 'maximum', at $\theta = 0.32$ degrees, is reduced to zero intensity.

The above example illustrates the general result that the *spacings* of the slits determine the angular *positions* of the diffraction maxima, whilst the *widths* of the slits determine how the *intensities* at the maxima decay with angle.

We conclude by deriving equation (3.3). The proof relies physically on Huygens' assumption that each portion of the wavefront emits secondary wavelets in all forward directions, and mathematically on the fact that

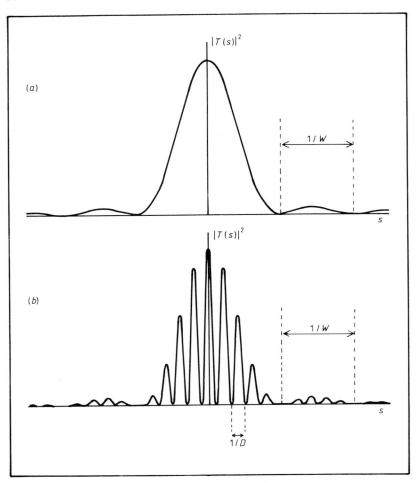

Figure 3.2 Graphs showing the angular distribution of diffracted intensity from (a) a single slit of width W, and (b) two slits each of width W with their centres a distance $D(=5W)$ apart. The variable s is approximately equal to θ/λ, when θ is small, and the intensity is proportional to $|T(s)|^2$.

Re$(e^{\pm i\beta}) = \cos \beta$, where Re means that we take the real part of the complex quantity in the bracket.

Consider the wavefront just emerging from the slit system, and suppose that the portion of width δx centred on $x=0$ emits Huygens wavelets which produce a wave disturbance at P whose dependence on time is $K \, \delta x \cos(2\pi f t)$. K is here a constant, depending on the incident intensity and on the distance R (at fixed λ), whilst f is the frequency of the radiation. We have chosen the time origin so as to give zero phase angle, for simplicity. Wavelets from other portions of the wavefront will, however, give a disturbance at P which is shifted in phase relative to that from $x=0$, because the distances of travel are different. We see from figure 3.1 that the distance from A to P is less than that from O to P by an amount $OB = x \sin \theta$, corresponding to a phase advance in radians of $2\pi x \lambda^{-1} \sin \theta$. This phase advance can be written $2\pi xs$, using $s = \lambda^{-1} \sin \theta$. The total disturbance at P, say F_P, is now obtained by adding up the increments from all wavelets, leading to the following integral. Note that the function $t(x)$ 'picks out' those portions of wavefront which get through the system of slits.

$$F_P = \int_{-\infty}^{+\infty} Kt(x) \cos(2\pi ft + 2\pi xs) \, dx$$

$$= K \int_{-\infty}^{+\infty} \mathrm{Re}\left[t(x) \exp(-i2\pi ft - i2\pi xs)\right] dx$$

$$= K \, \mathrm{Re}\left(\exp(-i2\pi ft) \int_{-\infty}^{+\infty} t(x) \exp(-i2\pi xs) \, dx\right)$$

$$= K \, \mathrm{Re}\left[T(s) \exp(-i2\pi ft)\right]$$

$$= K |T(s)| \cos(2\pi ft + \phi).$$

In this last expression ϕ depends on s and is defined so that $T(s) = |T(s)| \exp(-i\phi)$. Since the intensity of wavelike radiations is known to be proportional to the square of the oscillatory amplitude of the wave, it follows that $I(\theta) \propto |T(s)|^2$, thus completing the derivation of equation (3.3).

A warning is necessary, because Huygens' principle as used here is in fact only approximate, and is at its best for slits that are individually wide compared with λ; when this is the case most of the diffracted intensity is concentrated into small angles for which $\sin \theta \approx \theta$ and $s \approx \theta/\lambda$. When the slit widths are about equal to, or less than, λ our formula is useful as a rough approximation at moderate angles but fails badly for $\theta \gtrsim 45°$. Two assumptions thus lie behind the treatment given here; first that R is large compared with the total width of the slit system, and second that the individual slits are wide compared with the wavelength of the radiation. These conditions define what is often called Fraunhofer diffraction.

3.3 Oscillations on a String

The problem here is that of predicting the motion of a string, fixed at each end, when it is set vibrating by, for instance, being plucked. The shape of the vibrating string can change in a complicated way with time, although the original shape recurs at regular intervals of time, and our aim is to explain and describe this. Many important features of the motion can be explained using an approximate treatment based on the assumptions that the sideways displacements of the string are small and that the only physical parameters involved are the mass M of the string, its length L, and the tension τ in the string. The effects of stiffness and of frictional damping forces are thus ignored.

If the sideways displacement of the string at a distance x from one end is initially $y_0(x)$ and if the string is let go from rest, the problem becomes that of calculating the subsequent displacement $y(x, t)$ at position x at time t. When the above simplifying conditions are met it is a matter of observation (which can be explained using Newton's laws of motion) that whenever $y_0(x)$ has a sinusoidal dependence on x then all parts of the string will oscillate harmonically together at the same frequency. In such a case, if the spatial period (or wavelength) of $y_0(x)$ is λ then the corresponding frequency of oscillation is $f = c/\lambda$ where c is the speed of transverse waves along the string. The value of c is $(\tau L/M)^{1/2}$, a result also derived from Newton's laws and verified by experiment. Expressed mathematically, this result becomes

$$y(x, t) = A \sin(2\pi x/\lambda) \cos(2\pi c t/\lambda). \tag{3.4}$$

Note that the spatial term is a sine rather than cosine because the displacement is always zero at $x = 0$, whilst the temporal term is a cosine rather than sine because the displacement is a maximum at time zero when $y(x, 0) = y_0(x)$. The value of λ cannot be arbitrary because the displacement is zero at $x = L$ and this will only be so if L is equal to an integer number of half wavelengths, so that $L = n\lambda/2$. The motion associated with a positive integer n is called the *nth normal mode of oscillation*, and is described by

$$y(x, t) = A \sin(n\pi x/L) \cos(n\pi c t/L).$$

We now come to the use of Fourier methods to calculate the motion when $y_0(x)$ is not a sinusoidal function of x. Suppose that, for some specified set of amplitudes A_n,

$$y_0(x) = \sum_{n=1}^{\infty} A_n \sin(n\pi x/L). \tag{3.5}$$

It is natural to assume that each of the components in this sum will evolve in time, as would the corresponding normal mode on its own; and this assumption is found to be valid when the simplifying conditions described earlier are met. The resultant motion is thus

$$y(x, t) = \sum_{n=1}^{\infty} A_n \sin(n\pi x/L) \cos(n\pi c t/L). \tag{3.6}$$

The summation in equation (3.5) is, however, exactly the form of a half-range Fourier sine series, equation (2.3b), and it follows that *any* arbitrary initial profile $y_0(x)$ can be dealt with in this way (barring pathological local behaviour). The problem is thus solved. Given $y_0(x)$ we calculate the A_n according to Fourier's half-range sine formula, equation (2.4b),

$$A_n = (2/L) \int_0^L y_0(x) \sin(n\pi x/L)\,dx. \tag{3.7}$$

We then calculate $y(x,t)$ by substituting these values into equation (3.6). Note that since each normal mode in (3.6) will oscillate at a different frequency, the shape of the string will evolve in a complicated way with time. However, whenever t is a multiple of $2L/c$ the original shape is retrieved because the cosine term is then unity for all n.

Example 3.6 *The mid point of a string of length L, fixed at each end, is held a distance D to one side, the two halves of the string being straight; if the string is let go from rest, determine the profile of the string at time t when the wave speed on the string is c.* The initial profile is described by different equations for each half of the string; for $0 < x < L/2$ we have $y_0(x) = 2Dx/L$, whilst for $L/2 < x < L$ we have $y_0(x) = (2D/L)(L-x)$. Equation (3.7) gives

$$A_n = \int_0^{L/2} \frac{4Dx}{L^2} \sin(\pi n x/L)\,dx + \int_{L/2}^L \frac{4D(L-x)}{L^2} \sin(n\pi x/L)\,dx.$$

Writing these integrals as I_1 and I_2, integration by parts gives

$$I_1 = \left[\frac{-4Dx\cos(n\pi x/L)}{n\pi L}\right]_0^{L/2} + \int_0^{L/2} \frac{4D\cos(n\pi x/L)}{\pi n L}\,dx$$

$$= \frac{-2D\cos(n\pi/2)}{n\pi} + \frac{4D\sin(n\pi/2)}{n^2\pi^2}.$$

$$I_2 = \left[\frac{-4D(L-x)\cos(n\pi x/L)}{n\pi L}\right]_{L/2}^L - \int_{L/2}^L \frac{4D\cos(n\pi x/L)}{n\pi L}\,dx$$

$$= \frac{2D\cos(n\pi/2)}{n\pi} + \frac{4D\sin(n\pi/2)}{n^2\pi^2}.$$

$$A_n = (8D/n^2\pi^2)\sin(n\pi/2).$$

The sine term makes A_n zero for even values of n, whilst the odd terms alternate in sign. The answer is thus

$$y(x,t) = \sum_{n=1}^{\infty} \frac{8D}{n^2\pi^2} \sin(n\pi/2) \sin(n\pi x/L) \cos(n\pi c t/L).$$

The technique can be extended to allow for the initial transverse speed of the string being non-zero, say $y_0'(x)$. We replace equation (3.6) by

$$y(x,t) = \sum_{n=1}^{\infty} \sin(n\pi x/L)[A_n \cos(n\pi c t/L) + B_n \sin(n\pi c t/L)].$$

The A_n are given as before by equation (3.7), but we now have the additional initial condition,

$$y'_0(x) = \frac{\partial y(x,t)}{\partial t}\bigg|_{t=0} = \sum_{n=1}^{\infty} \frac{nB_n \pi c}{L} \sin(n\pi x/L)$$

so that the Fourier formula for this half-range series gives B_n through

$$\frac{nB_n \pi c}{L} = \frac{2}{L} \int_0^L y'_0(x) \sin(n\pi x/L)\,dx.$$

3.4 Heat Conduction

An object whose parts are at different temperatures will, in the course of time, settle to a uniform temperature provided there is no flow of heat in or out of the object. In this section, we show how Fourier methods are useful in understanding the way such temperature distributions change with time.

We consider one of the simplest such systems, a uniform rod of length L and cross-sectional area A which is thermally insulated from its surroundings. Since no heat is allowed in or out, the average temperature of the whole rod will not change with time, and it is convenient to deal throughout with temperatures measured relative to this average. So, at time $t=0$, let the temperature at a distance x from one end be equal to the average temperature plus $\theta_0(x)$, in degrees Celsius or degrees Kelvin. $\theta_0(x)$ will be positive for some values of x and negative for others. If correspondingly the temperature at position x at time t is equal to the average temperature plus $\theta(x,t)$, our aim is to calculate $\theta(x,t)$ from $\theta_0(x)$. We assume that temperatures across the rod, at constant x, are uniform so that no further parameters enter the calculation except for those describing the material of the rod.

An initial distribution $\theta_0(x)$ proportional to $\cos(\pi x n/L)$, for any positive integer n, is special because we can show in this case that the temperature fluctuations simply die away exponentially with time. Thus in this special case

$$\theta(x,t) \propto \exp(-\mu_n t)\cos(\pi n x/L) \tag{3.8}$$

where the constant μ_n depends on the index n and on the thermal conductivity λ, specific heat capacity C, and the density ρ of the material of the rod as follows:

$$\mu_n = \lambda \pi^2 n^2 / (C\rho L^2). \tag{3.9}$$

We justify (3.8) and (3.9) later, but now proceed to the case when $\theta_0(x)$ is not cosinusoidal.

Since $\theta_0(x)$ is only defined for $0 \leq x \leq L$ we can expand it as a half-range Fourier cosine series

$$\theta_0(x) = \sum_{n=1}^{\infty} A_n \cos(\pi n x/L) \tag{3.10}$$

where A_n is calculated from the Fourier formula, equation (2.4a),

$$A_n = \frac{2}{L} \int_0^L \theta_0(x) \cos(\pi n x/L) \, dx. \qquad (3.11)$$

Assuming now that each of the cosine contributions in the summation in (3.10) evolves with time as in (3.8), we obtain as the solution to our problem

$$\theta(x,t) = \sum_{n=1}^{\infty} A_n \exp\left(-\frac{\lambda \pi^2 n^2 t}{C \rho L^2}\right) \cos(n\pi x/L) \qquad (3.12)$$

where the A_n are obtained from $\theta_0(x)$ as in equation (3.11).

Example 3.7 *An insulated copper rod of length 0.1 m has, initially, one half at a uniform temperature 2 K higher than the other half, also at uniform temperature. Find the approximate temperature distribution along the rod after 2 s using the values* $\lambda = 383$ W m^{-1} K^{-1}, $C = 385$ J kg^{-1} K^{-1}, $\rho = 8.89 \times 10^3$ kg m^{-3}. $\theta_0(x)$ has the value -1 K for $0 \leq x < (L/2)$, and the value $+1$ K for $(L/2) < x < L$. Thus equation (3.11) becomes

$$A_n = \frac{2}{L} \int_0^{L/2} \cos\left(\frac{\pi n x}{L}\right) dx - \frac{2}{L} \int_{L/2}^L \cos\left(\frac{\pi n x}{L}\right) dx$$

$$= (4/\pi n) \sin(n\pi/2) \qquad (n = 1, 2, 3, \ldots).$$

A_n is zero for even n, and the first few non-zero coefficients are $A_1 = 1.273$ K, $A_3 = -0.424$ K, $A_5 = 0.255$ K. At time t we replace the coefficients A_n by the modified values $A'_n = A_n \exp(-\mu_n t)$, with μ_n as in (3.9). With the numbers given we obtain, for $t = 2$ s: $A'_1 = 1.022$ K, $A'_3 = -0.059$ K, and $A'_5 = 0.001$ K. Clearly, we can as a good approximation retain only A'_1 and A'_3, so that (3.12) gives finally as our answer, in units of kelvin at $t = 2$ s, with x in metres and the trigonometric arguments in radians,

$$\theta(x,t) \simeq 1.02 \cos(10\pi x) - 0.06 \cos(30\pi x).$$

We now conclude by showing that equations (3.8) and (3.9) are indeed consistent with the physical laws of heat flow. Let us write the temperature gradient at an arbitrary value $x = X$ as $(\partial \theta/\partial x)_{x=X}$; this is obtained from $\theta(x,t)$ by differentiating with respect to x, at fixed t, and then substituting $x = X$ in the resulting expression. The law of heat conduction now tells us that the heat passing per unit time at $x = X$ into the portion of rod extending from $x = 0$ to $x = X$ is $A\lambda(\partial \theta/\partial x)_{x=X}$. This flow of heat must be equal to the product of the total heat capacity of the portion from $x = 0$ to $x = X$ multiplied by the rate of increase of its average temperature. Bearing in mind that this heat capacity is $\rho C A X$, we have the equation

$$A\lambda(\partial \theta/\partial x)_{x=X} = \rho C A X \frac{d}{dt}\left(\frac{1}{X} \int_0^X \theta(x,t) \, dx\right). \qquad (3.13)$$

A function $\theta(x,t)$ will only be physically acceptable if it satisfies this equation. On substituting the trial function of (3.8) into (3.13), we find after performing the various differentiations and integrations that (3.13) is satisfied only if μ_n has the value in (3.9). Note that a cosine function rather than a sine function is used throughout because it is necessary that $(\partial\theta/\partial x)_{x=0}$ should be zero to satisfy the boundary condition of zero heat flow at $x=0$. Further, the argument of the cosine is chosen as $(\pi n x/L)$ in order that $(\partial\theta/\partial x)_{x=L}$ should be zero to satisfy the other boundary condition of zero heat flow at $x=L$.

3.5 The Bandwidth Theorem and Uncertainty Principle

In many applications it is useful to know how the duration of a transient function of time, $y(t)$, is related to the range of frequencies that are present in its transform, $Y(f)$. One finds that the shorter the transient the larger is the spread of frequencies in its transform. In electrical engineering, these ideas find mathematical expression in the *bandwidth theorem*. There is analogously a relation between the length of a spatial transient, $y(x)$, and the range of spatial frequencies covered by its transform, $Y(s)$. In the theory of wave mechanics, an application of these ideas to a wave function $\psi(x)$ leads, as we shall see, to a formulation of the famous *uncertainty principle*.

If we use Δt to represent the duration of the transient $y(t)$, and if Δf represents the range of frequencies spanned by $Y(f)$, that is its *bandwidth*, then we find for several simple functions $y(t)$ that

$$\Delta t \, \Delta f \simeq 1. \qquad (3.14)$$

This is the *bandwidth formula*. The approximate sign means here that the product usually lies in the range 0.5–3 for transients with simple graphs. Since $y(t)$ and $Y(f)$ often decay only slowly to zero as t or f tend to infinity, it is not always clear how actually to define Δt and Δf; the above generalisation rests on using 'natural' definitions as in the following examples. We give a more precise approach later.

Example 3.8 *Estimate the bandwidth product, $\Delta t \, \Delta f$, for the transient $y(t) = 1$ $(-T/2 < t < T/2)$, $= 0$ ($|t| > T/2$).* This function and its transform $Y(f)$ have already been met in example 2.6 and figure 2.3(d). It is natural to define $\Delta t = T$. A natural definition for Δf is less obvious, but one simple method is to put it equal to the frequency difference between the minima on either side of the main maximum in the graph of $Y(f)$. Given that $Y(f)$ is equal to $(\pi f)^{-1} \sin(\pi f \Delta t)$, these minima occur at $f = \pm \Delta t^{-1}$, so that $\Delta f = 2/\Delta t$. As a result, the bandwidth product $\Delta t \, \Delta f$ has the value 2.

Note that as Δt gets smaller so Δf gets larger. An example shows a simple but important application. If a cable is to transmit an electric pulse of duration $\Delta t = 1$ μs satisfactorily, it is necessary that the cable should be able to carry frequencies up to at least $\Delta f/2$, that is up to 1 MHz.

Example 3.9 *Estimate the bandwidth product for the so-called gaussian function, defined for all t as equal to $\exp(-a^2 t^2)$ where a is a positive constant.* The graph of this function consists of a smooth hump centred on the origin; its value at the origin is unity, and its value is $1/e$ when $t = \pm 1/a$. A natural and simple estimate of the duration of this pulse is $\Delta t = 2/a$. The transform can be obtained by integration, but it is a lengthy calculation and we therefore quote the result:

$$Y(f) = \pi^{1/2} a^{-1} \exp(-\pi^2 f^2/a^2).$$

This also is a gaussian function, and $Y(f)$ is equal to its peak value divided by e when $f = \pm a/\pi$. We thus put $\Delta f = 2a/\pi$, and the bandwidth product is $\Delta t \, \Delta f = 4/\pi$.

When we consider a spatial transient $y(x)$ with transform $Y(s)$, where s represents spatial frequency, then the analogue of the bandwidth formula is

$$\Delta x \, \Delta s \simeq 1$$

where Δx represents the length of $y(x)$, and Δs represents the spread of spatial frequencies covered by $Y(s)$. Let us see the implications of this when applied to the Schrödinger wave function $\psi(x)$ of a particle moving in one dimension. When $\psi(x)$ is a spatial transient we usually call it a wave packet. Δx now measures the length of the wave packet, and Δs measures the range of spatial frequencies in the transform, $Y(s)$, of $\psi(x)$. However, the variable s, equal to inverse wavelength, has another interpretation. The de Broglie relation for particle-waves relates the momentum p of a particle to its wavelength λ by the formula $p = h/\lambda$, where $h \simeq 6.63 \times 10^{-34}$ J s is Planck's constant. This formula receives its most direct confirmation in experiments on electron diffraction. Thus $p = hs$, and the parameter Δs is related to an equivalent spread of momenta Δp, where $\Delta p = h \, \Delta s$. The bandwidth formula is now equivalent to

$$\Delta x \, \Delta p \simeq h \qquad (3.15)$$

where Δx is the length of a wave packet and Δp is the range of momenta associated with it. It is a tenet of wave mechanics that $|\psi(x)|^2 \, \delta x$ is equal to the probability of the particle being observed at a position in the range x to $x + \delta x$, so that Δx is a measure of the uncertainty in our knowledge of the particle's position. It is likewise a tenet that $|Y(s)|^2 \, \delta s$ is equal to the probability of the particle being observed to have a momentum in the range hs to $h(s + \delta s)$, so that Δp is a measure of the uncertainty in our knowledge of the particle's momentum. Equation (3.15) is now an approximate version of the *uncertainty principle*.

The above discussions are marred by the lack of a single and precise method of defining Δt, Δf, Δx, Δs and Δp. A widely used *precise* method of definition is as follows:

$$(\Delta t)^2 = \int_{-\infty}^{+\infty} t^2 |y(t)|^2 \, dt \Big/ \int_{-\infty}^{+\infty} |y(t)|^2 \, dt$$

$$(\Delta f)^2 = \int_{-\infty}^{+\infty} f^2 |Y(f)|^2 \, df \Big/ \int_{-\infty}^{+\infty} |Y(f)|^2 \, df.$$

These integrals are often not easy to evaluate, but they have the merit of giving a single prescription applicable to a wide variety of pulse shapes. The analogous definitions of Δx, Δs and thence Δp follow naturally. These definitions have the further merit that whenever they lead to finite values of Δt and Δf it is possible to establish rigorously that

$$\Delta t \, \Delta f \geqslant (4\pi)^{-1}. \tag{3.16}$$

This inequality is the precise statement of the bandwidth theorem. The corresponding precise statement of the uncertainty principle is

$$\Delta x \, \Delta p \geqslant (h/4\pi). \tag{3.17}$$

It can be shown that the gaussian function is the only one leading to an equality in equations (3.16) and (3.17); all other functions yield a larger product.

Further Topics

4.1 Sampling and Computation

The examples of Fourier analysis given so far have been restricted to rather artificially simple functions, in order that the resulting Fourier integrals should be easy to evaluate. The Fourier equations remain valid for more complicated functions, but then we often need a numerical technique to evaluate the resulting integrals. The numerical techniques are ideally suited to computer calculation, and we now describe them.

Consider the periodic function $y(t)$ whose graph appears in figure 4.1(a). This graph is based on a tracing of the profile of a face drawn by Leonardo da Vinci (rotate the page 90° anticlockwise to see this). It is clear that $y(t)$ is not a simple mathematical function and the integrals needed to evaluate the Fourier series coefficients will not be obtainable using standard integrals. Instead, we show how to obtain the integrals by interpreting them as areas, using numerical techniques. Consider the coefficient A_0,

$$A_0 = \frac{2}{T} \int_0^T y(t)\,dt. \tag{4.1}$$

To approximate this integral we divide the graph of one period of $y(t)$ into a large number, P, of strips as in figure 4.1(b), drawn with $P=16$ for simplicity. If the height of the nth vertical (at $t = nT/P$) is written as y_n, for $n = 0, 1, 2, \ldots, P-1$, then the area of the nth strip is approximately $y_n T/P$. Equation (4.1) now becomes as below, on adding the areas of all strips together,

$$A_0 \simeq (2/T) \sum_{n=0}^{P-1} (y_n T/P). \tag{4.2}$$

Clearly, as $P \to \infty$ so this approximation becomes more exact. The process of obtaining the set of values y_n at regular spaced intervals is called *sampling* the function $y(t)$, and the y_n are called the sampled values of $y(t)$. The integral for the mth Fourier cosine coefficient is evaluated in a similar way, by sampling the whole integrand at times $t = nT/P$, including the cosine term; thus

40 *Fourier Transforms in Physics*

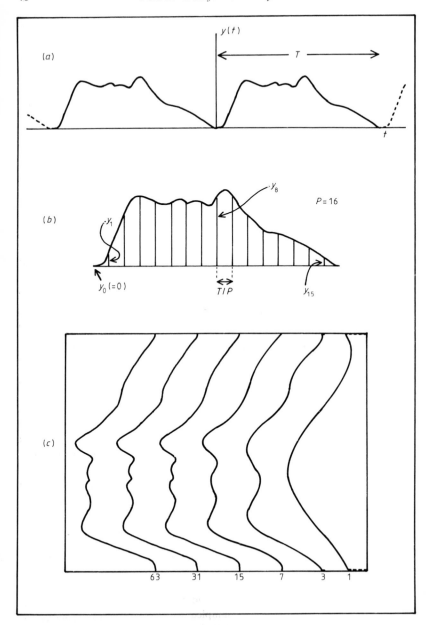

Further Topics

$$A_m = (2/T) \int_0^T y(t) \cos(2\pi mt/T) \, dt$$

$$\simeq (2/T) \sum_{n=0}^{P-1} [y_n \cos(2\pi mn/P)(T/P)]. \tag{4.3}$$

Similarly the sine coefficients are approximated by

$$B_m \simeq (2/P) \sum_{n=0}^{P-1} [y_n \sin(2\pi mn/P)]. \tag{4.4}$$

In each of these summations we fix the integer m, and the summation runs over the indicated values of n.

The process of sampling means that fine structure in $y(t)$ is 'lost'; the result of this is that the approximate values for A_m and B_m given above become worse as larger values of m are used. As a very rough rule, the approximation will be of little use unless $m \leqslant P/8$.

Example 4.1 *Obtain approximate values for the Fourier series coefficients* A_0, A_1 *and* B_1 *of the 'Leonardo' function of figure 4.1(a), given that the sampled values of $y(t)$ at times $t = nT/P$, for $n = 0$ to 7 and $P = 8$, are:* 0, 2.65, 3.32, 3.31, 3.55, 2.65, 1.69, 0.98. A pocket calculator is sufficient to give us, from equation (4.2),

$$A_0 \simeq (2/8)(0 + 2.65 + \ldots + 0.98)$$
$$\simeq 4.54.$$

Likewise A_1 and B_1 can be obtained from equations (4.3), (4.4) with $m = 1$, and a pocket calculator gives

$$A_1 \simeq (2/8)[0 + 2.65 \cos(2\pi/8) + 3.32 \cos(4\pi/8) + \ldots]$$
$$\simeq -1.30$$

$$B_1 \simeq (2/8)[0 + 2.65 \sin(2\pi/8) + 3.32 \sin(4\pi/8) + \ldots]$$
$$\simeq +0.82.$$

These may be compared with 'true' values, calculated using 128 sample points, of $A_0 = 4.56$, $A_1 = -1.34$, $B_1 = +0.82$.

The Fourier transform of a function $y(t)$ can be approximated in a similar way, except that we have first to *truncate* $y(t)$ before sampling it. This means that we replace $y(t)$ by zero when $|t| > T/2$, where T is some arbitrary but fixed value of t; in order that errors arising from truncation should be small we must choose T to be large enough. If now y_n is the sampled value of $y(t)$ at $t = nT/P$, with the integer running from $-P/2$ up through zero to $+(P/2) - 1$, where P is a large even

◀ **Figure 4.1** (a) and (b) Sampling a periodic function that is based on a profile by Leonardo da Vinci. (c) Reconstruction of the profile using 128 sample points for the analysis, but only 63, or 31, or 15, etc harmonics, as labelled, in the synthesis: most of the significant detail is contained in the first 31 harmonics.

integer, then the Fourier transform $Y(f)$ of $y(t)$ becomes

$$Y(f) \simeq \int_{-T/2}^{T/2} y(t) \exp(-2\pi i f t) \, dt$$
$$\simeq \sum_{n=-P/2}^{P/2-1} [y_n \exp(-2\pi i f n T/P)(T/P)]. \qquad (4.5)$$

To construct the function $Y(f)$ this summation has to be evaluated at many different values of f.

This approximate version of $Y(f)$ is subject to two types of error which can occur individually or together. There is a truncation error if T is too small, and a sampling error if N/T is too small. The truncation error, on its own, means that whilst the general shape of the graph of $Y(f)$ is correct, the local structure is falsified, there being a rounding off of any sharp features in the graph. The sampling error, on its own, means that $Y(f)$ becomes less accurate as $|f|$ increases, and the approximation is only useful if $|f| \ll P/T$. In view of these errors there is little point in calculating $Y(f)$ at too many values of f, and a set of values $f = m/T$ is convenient, with the integer m running from $-P/2$ up through zero to $(P/2) - 1$. The value of $Y(f)$ at $f = m/T$ is thus

$$Y(m/T) \simeq (T/P) \sum_{n=-P/2}^{+P/2-1} y_n \exp(-2\pi i m n/P) \qquad (4.6)$$

the approximation being best when $|m| \ll P$.

4.2 The Fast Fourier Transform

Summations such as those in the previous section are well adapted to computer calculation using BASIC or FORTRAN languages. For example to find the Fourier cosine and sine coefficients, A(M) and B(M), from the P sampled values, Y(N), of a periodic function using equations (4.3), (4.4), the core of a BASIC program might be as follows:

```
 90   DIM A(P),B(P), Y(P)
100   FOR M=0 TO P−1 : A(M) = 0 : B(M) = 0
110   FOR N=0 TO P−1
120   A(M)=A(M) + Y(N)*COS(2*PI*M*N/P)
130   B(M)=B(M) + Y(N)*SIN(2*PI*M*N/P)
140   NEXT N
150   A(M)=A(M)/P : B(M)=B(M)/P
160   NEXT M
```

This will be suitable for use on most microcomputers if combined with inputs setting up the values of P and Y(N), for $N = 0$ to $P - 1$, and followed by stages for printing out the arrays A(M) and B(M).

However, it will be found that for values of P greater than about 32 such a

Further Topics

program can take several minutes to run, or several hours for $P \gtrsim 256$. This arises because the operations in lines 120 and 130 have to be carried out P^2 times. The so-called *fast Fourier transform* (FFT) is a way of laying out the computation which is much faster for large values of P; it is simply an ingenious way of achieving the same result as above but using fewer additions and multiplications. The use of the FFT in computing was introduced by J W Cooley and J W Tukey in 1964 and its use has revolutionised instrumentation based on Fourier techniques. We reproduce below a program in BASIC, suitable for use on BBC machines, which a reader can copy without the need to understand its operations in detail. With minor modifications, it will be suitable for Apple machines.

Given an array of (perhaps complex) numbers g_n ($n = 0, 1, \ldots, P-1$), the program will compute the numbers G_m, ($m = 0, 1, \ldots, P-1$) where

$$G_m = \sum_{n=0}^{P-1} g_n \exp(-2\pi i m n / P). \tag{4.7}$$

To use the program, P must be assigned a value such as $2, 4, 8, 16, \ldots$, which is a power of 2, say 2^S. The real and imaginary parts of the g_n are assigned to the arrays RE(N) and IM(N), with $n = N$. The program runs through various stages, each of which reassigns the contents of these arrays until on completion the numbers RE(M) and IM(M) are equal to the real and imaginary parts of the G_m, with $m = M$. The first stage is known as 'bit reversal' and there are then S further stages, the progress being printed on the screen.

```
10  REM FAST FOURIER TRANSFORM
20  INPUT "GIVE NUMBER OF POINTS,P",P% : S%=INT(LN(P%)/LN(2)
    +0.999) : CLS
30  R1%=2^S%:R%=R1%-1:R2%=R1%DIV2:R4%=R1%DIV4:
    R3%=R4%+R2%
40  DIM RE(R1%),IM(R1%),CO(R3%)
50  FOR N%=0 TO P%-1 : PRINT "POINT NUMBER, N = ";N%
60  INPUT "RE(N)",RE(N%),"IM(N)",IM(N%)
70  NEXT
80  S2%=S% DIV 2 : S1%=S%-S2%
90  P1%=2^S1%:P2%=2^S2%
100 DIM V%(P1%-1) : V%(0)=0 : DV%=1 : DP%=P1%
110 FOR J%=1 TO S1% : HA%=DP% DIV 2 : PT%=P1%-HA%
120 FOR I%=HA% TO PT% STEP DP% : V%(I%)=V%(I%-HA%)+DV% :
    NEXT
130 DV%=DV%+DV% : DP%=HA% : NEXT
140 K=2*PI/R1%
150 FOR X%=0 TO R4% : COX=COS(K*X%) : CO(X%)=COX
160 CO(R2%-X%)=-COX : CO(R2%+X%)=-COX : NEXT
170 PRINT "FFT : BIT REVERSAL"
180 FOR I%=0 TO P1%-1 : IP%=I%*P2%
190 FOR J%=0 TO P2%-1 : H%=IP%+J% : G%=V%(J%)*P2%+V%(I%)
```

```
200  IF G%>H% TEMP=RE(G%) : RE(G%)=RE(H%) : RE(H%)=TEMP
210  IF G%>H% TEMP=IM(G%) : IM(G%)=IM(H%) : IM(H%)=TEMP
220  NEXT : NEXT
230  T%=1
240  FOR ST%=0 TO S%-1 : PRINT "STAGE ";ST% : D%=R2% DIV T%
250  FOR Z%=0 TO T%-1 : L%=D%*Z% : LS%=L%+R4%
260  FOR I%=0 TO D%-1 : A%=2*I%*T%+Z% : B%=A%+T% : F1=RE(A%)
     : F2=IM(A%)
270  P1=CO(L%)*RE(B%) : P2=CO(LS%)*IM(B%) : P3=CO(LS%)*RE(B%)
     : P4=CO(L%)*IM(B%)
280  RE(A%)=F1+P1-P2 : IM(A%)=F2+P3+P4 : RE(B%)=F1-P1+P2 :
     IM(B%)=F2-P3-P4
290  NEXT : NEXT : T%=T%+T% : NEXT
300  CLS : PRINT; "M";TAB(4);"RE(M)";TAB(20);"IM(M)"
310  FOR M%=0 TO R% : PRINT;M%;TAB(4);RE(M%);TAB(20);IM(M%) :
     NEXT
320  END
```

Example 4.2 *Given $g_0=2, g_1=0, g_2=1, g_3=0$, determine the values of G_m from equation (4.7), with $P=4$.* On running the FFT program we type in 4, as invited, for the value of P, and type in the values 2, 0, 0, 0, 1, 0, 0, 0, when requested, for the values of RE(0), IM(0), RE(1), IM(1), etc. On output, we obtain the values 3, 1, 3, 1, for the real parts of G_0 to G_3, and zero for the imaginary parts (to within the computer's accuracy).

The transformation from the g_n to the G_m in equation (4.7) is called the *discrete Fourier transform* (DFT), and its inverse is given exactly by

$$g_n = (1/P) \sum_{m=0}^{P-1} G_m \exp(+2\pi i mn/P). \tag{4.8}$$

The program can be modified to compute this inverse DFT by inserting the lines

```
65   RE(N%) = RE(N%)/P% : IM(N%) = IM(N%)/P%
275  P2 = -P2 : P3 = -P3
```

respectively after lines 60 and 270.

Example 4.3 *Following the previous example, show that the inverse DFT of the set $G_0=3, G_1=1, G_2=3, G_3=1$ yields back the original set, $g_0=2, g_1=0, g_2=1, g_3=0$.* We insert lines 65 and 275 into the program, and then type in the values 3, 0, 1, 0, 3, 0, 1, 0, as invited, for the values of RE(0), IM(0), RE(1), IM(1) etc. On output we obtain the values of the gs.

For Fourier series analysis we use the FFT program as follows, after inserting line 65 but not line 275. Given $y(t)$, of period T, choose P as equal to some power

Further Topics 45

of 2 and obtain the sampled values $y(nT/P)$ with the integer n running from 0 to $P-1$. Then assign RE(N) and IM(N) so that, with $n=N$,

$$y(nT/P) = \text{RE(N)} + i\text{IM(N)}.$$

The output will yield the complex Fourier series coefficients Y_n, and thence the cosine and sine coefficients (equations (2.11), (2.10)), as follows with $n=M$:

$$Y_n \simeq \text{RE(M)} + i\text{IM(M)} \qquad 0 \leqslant n \leqslant P/2$$

$$Y_{-n} \simeq \text{RE}(P-M) + i\text{IM}(P-M) \qquad 1 \leqslant n \leqslant P/2.$$

The approximation will be best for $|n| \ll P/2$.

For the evaluation of Fourier transforms we use the FFT program as follows, with line 65 inserted but not line 275. Given $y(t)$ we choose the integer P, equal to some power of 2, and choose the truncation parameter T. The sampled values $y(nT/P)$, with n running from $-P/2$ through zero to $(P/2)-1$ are used to assign RE(N) and IM(N) according to

$$y(nT/P) = \text{RE(N)} + i\text{IM(N)}$$

where for $n \geqslant 0$ we put $n = N$, and for $n < 0$ we put $N = P - |n|$. On output we obtain an approximation to $Y(f)$, equation (4.5), at $f = m/T$ with m running from $-P/2$ to $(P/2)-1$, using

$$(1/T)Y(m/T) \simeq \text{RE(M)} + i\text{IM(M)}.$$

For $m \geqslant 0$ we put $M = m$, and for $m < 0$ we put $M = P - |m|$. The approximation will be best for small values of $|m|$.

For Fourier series synthesis we use the FFT program as follows, with line 275 inserted but not line 65. Given a set of complex Fourier coefficients Y_n, we wish to evaluate

$$y(t) = \sum_{n=-H}^{H} Y_n \exp(2\pi i n t/T)$$

for a given period T and a given value of H, the harmonic number of the highest frequency to be included. To calculate $y(t)$ at a set of $P(=2^S > 2H)$ points spread over one period, we assign RE(N) and IM(N) as follows, with $n = N$:

$$Y_n = \text{RE(N)} + i\text{IM(N)} \qquad 0 \leqslant n \leqslant H$$

$$Y_{-n} = \text{RE}(P-N) + i\text{IM}(P-N) \qquad 0 < n \leqslant H. \qquad (4.9)$$

The remaining members of the arrays RE(N) and IM(N), that is from $N = H+1$ up to $N = P-H-1$, are assigned the value zero. On output the values of the synthesised function $y(t)$, at $t = (mT/P)$ for m running from 0 to $P-1$, are given by the following with $m = M$,

$$y(mT/P) = \text{RE(M)} + i\text{IM(M)}.$$

Since no integration is involved the result is exact. For the special case that the Fourier cosine and sine coefficients A_n and B_n are all real, equation (4.9) reduces to the following, with $n = N$:

$$RE(0) = A_0/2, \quad IM(0) = 0$$
$$RE(N) = RE(P - N) = A_n/2 \quad\quad 1 \leqslant n \leqslant H$$
$$IM(N) = -IM(P - N) = -B_n/2 \quad\quad 1 \leqslant n \leqslant H$$

whilst the remaining members of RE(N) and IM(N) are equal to zero.

4.3 Rigorous Fourier Theorems

We now indicate how the Fourier formulae are justified. Truly rigorous proofs are lengthy, and we will give condensed versions, followed by a discussion of the assumptions involved. We start with the formulae for the Fourier series of coefficients.

Given a real or complex valued periodic function $y(t)$ of period T, let us assume that coefficients A_n and B_n exist such that

$$y(t) = A_0/2 + \sum_{n=1}^{\infty} [A_n C_n(t) + B_n S_n(t)] \quad (4.10)$$

where we have used the abbreviations

$$C_n(t) = \cos(2\pi nt/T) \quad\quad S_n(t) = \sin(2\pi nt/T).$$

Our aim is to derive the Fourier formulae for the A_n and B_n. It is convenient to introduce the further abbreviation $H_n(t)$ for the nth harmonic, that is:

$$H_n(t) = A_n C_n(t) + B_n S_n(t) \quad\quad n \geqslant 1$$
$$H_0(t) = A_0/2.$$

Standard methods of integration lead to the following results, for any pair of integers $m, n \geqslant 0$; the values depend only on whether $m = n$ or not:

$$\int_0^T H_n(t) C_m(t) \, dt = \begin{cases} A_m T/2 & m = n \\ 0 & m \neq n \end{cases} \quad (4.11)$$

$$\int_0^T H_n(t) S_m(t) \, dt = \begin{cases} B_m T/2 & m = n \\ 0 & m \neq n. \end{cases} \quad (4.12)$$

The main 'proof' now runs as follows. To find a particular coefficient, say A_m, we multiply both sides of equation (4.10) by $C_m(t)$ and then integrate each side over one period of t. Noting that m remains fixed whilst the summations run over n we obtain:

$$\int_0^T y(t) C_m(t) \, dt = \int_0^T \sum_{n=0}^{\infty} H_n(t) C_m(t) \, dt \quad (4.13)$$

$$= \sum_{n=0}^{\infty} \int_0^T H_n(t) C_m(t) \, dt$$

$$= 0 + 0 + \ldots + A_m T/2 + 0 + 0 + \ldots$$

This is the required formula giving A_m, and on replacing $C_m(t)$ throughout by $S_m(t)$ we can obtain the corresponding formula for B_m. The string of zeros arises from using equations (4.11), (4.12); only one term in the infinite summation survives.

The three key assumptions in the above derivation are (i) that the expansion in equation (4.10) is possible, (ii) that $y(t)C_m(t)$ is integrable over one period, and (iii) that the order of the integration and summation in equation (4.13) can be reversed. Since Fourier's work, various sets of conditions have been discovered which will ensure that these assumptions are justified. One set is as follows. If the periodic function $y(t)$ is defined and non-infinite at all t and if, over any finite interval of t, $y(t)$ is such that: (i) there are only a finite number of discontinuities, (ii) there are only a finite number of maxima and minima, and (iii) at a discontinuity $y(t)$ has a value half way between the values on each side of the discontinuity, then (4.10) will be valid at *all* t and the coefficients will be given by the Fourier formulae. We call these the Dirichlet conditions, after their discoverer. If $y(t)$ is complex then the conditions are applied to the real and imaginary parts separately.

Now consider the Fourier formula for transforms. Given a non-periodic function $y(t)$ we assume that a function $Y(f)$ can be defined by

$$Y(f) = \int_{-\infty}^{+\infty} y(t) \exp(-2\pi i f t) \, dt \tag{4.14}$$

and our aim is to show that

$$y(t) = \int_{-\infty}^{+\infty} Y(f) \exp(+2\pi i f t) \, df. \tag{4.15}$$

It is useful, as a preliminary, to establish the values of two integrals. First,

$$\int_{-\infty}^{+\infty} e^{-|af|} \exp[2\pi i f(t-u)] \, df = \frac{2a}{a^2 + 4\pi^2(u-t)^2}.$$

This is simply a variant of example 2.5, using different symbols. Next we need to establish that the quantity on the right-hand side of this expression integrates to a value of unity when integrated over all values of u, ($-\infty$ to $+\infty$), at constant t and a. This integration reduces to a standard form if the substitution $x = (2\pi/a)(u-t)$ is used. The main 'proof' now rests upon considering the value of the following repeated integral,

$$I = \int_{-\infty}^{-\infty} \int_{-\infty}^{+\infty} y(u) e^{-|af|} \exp[2\pi i f(t-u)] \, df \, du$$

and assuming that its value is independent of the order in which the integrations are performed. Integrating over u first (at constant f), we obtain, using equation (4.14),

$$I = \int_{-\infty}^{+\infty} Y(f) e^{-|af|} \exp(2\pi i f t) \, df \qquad (4.16)$$

whilst integrating over f first (at constant u) gives:

$$I = \int_{-\infty}^{+\infty} y(u) \left(\frac{2a}{a^2 + 4\pi^2(u-t)^2} \right) du. \qquad (4.17)$$

On taking the limits as $a \to 0$ in each expression for I and equating the results, the required formula for $y(t)$ emerges. Note that as $a \to 0$ so $e^{-|af|} \to 1$. Note also that as $a \to 0$ so the bracketed quantity in equation (4.17) becomes very small except for values of u such that $u \simeq t$; $y(u)$ can thus be replaced by $y(t)$ and treated as a constant so far as the integration is concerned. Thus, using (4.16) and (4.17) respectively we obtain

$$\lim_{a \to 0} I = \int_{-\infty}^{+\infty} Y(f) \exp(2\pi i f t) \, dt$$

$$= \lim_{a \to 0} y(t) \int_{-\infty}^{+\infty} \left(\frac{2a}{a^2 + 4\pi^2(u-t)^2} \right) du$$

$$= y(t).$$

This is the desired result.

The key assumptions in the above derivation are (i) that the various integrals do exist, (ii) that the value of I is independent of the order of integration, and (iii) that the steps based on the limit as $a \to 0$ are valid. If throughout we interpret an integral over infinite limits in the sense that $\int_{-\infty}^{+\infty} = \lim_{X \to \infty} \int_{-X}^{+X}$ (the so-called Riemann–Cauchy limit), then the above assumptions will be satisfied if $y(t)$ is everywhere defined and non-infinite, and if also $y(t)$ satisfies the Dirichlet conditions over every finite interval of t, and if in addition $\int_{-\infty}^{+\infty} |y(t)| \, dt$ is finite. Equations (4.14) and (4.15) will then be valid for *every* value of f and t respectively. This is by no means the only set of sufficient conditions that is known, and each such set constitutes a *Fourier inversion theorem*.

The Parseval formula for a function $y(t)$, having Fourier transform $Y(f)$ as in equation (4.14), may be derived as follows. Consider the repeated integral

$$I = \int_{-\infty}^{+\infty} \int_{-\infty}^{+\infty} y(t) Y^*(f) \exp(-2\pi i f t) \, df \, dt.$$

On assuming that the order of integration is immaterial we obtain, on integration over f and t respectively and using equations (4.14) and (4.15),

$$\int_{-\infty}^{+\infty} y(t) y^*(t) \, dt = \int_{-\infty}^{+\infty} Y(f) Y^*(f) \, df$$

as required. Once again there are several sets of conditions which will ensure the validity of this result. One condition is simply that $\int_{-\infty}^{+\infty} |y(t)|\, dt$ and $\int_{-\infty}^{+\infty} |y(t)|^2\, dt$ shall be finite, without the need for the Dirichlet conditions to be satisfied.

The Parseval formula for Fourier series, equation (2.27), is likewise valid provided simply that $\int_0^T |y(t)|\, dt$ and $\int_0^T |y(t)|^2\, dt$ are finite.

In summary, for most elementary practical applications of Fourier analysis one can take it for granted that the Fourier and Parseval formulae will be valid.

Appendix 1: Some Fourier Series

We list below some Fourier series of period $T(=2\pi/\omega)$ whose sums are given for values of t within the stated ranges: values at other values of t are obtained using the periodicity property. At a value of t corresponding to a discontinuity the series sums to a value midway between the limiting values each side of the discontinuity. For each of the four pairs of series, the second series results from delaying its partner by one quarter of a period.

Square Waveforms

$$(4/\pi)[\cos \omega t - (\cos 3\omega t)/3 + (\cos 5\omega t)/5 - \ldots] = \begin{cases} 1 & -T/4 < t < T/4 \\ -1 & T/4 < t < 3T/4 \end{cases}$$

$$(4/\pi)[\sin \omega t + (\sin 3\omega t)/3 + (\sin 5\omega t)/5 + \ldots] = \begin{cases} 1 & 0 < t < T/2 \\ -1 & T/2 < t < T. \end{cases}$$

Triangular Waveforms

$$(8/\pi^2)[\cos \omega t + (\cos 3\omega t)/3^2 + (\cos 5\omega t)/5^2 + \ldots] = 1 - 4|t|/T \qquad -T/2 \leqslant t \leqslant T/2$$

$$(8/\pi^2)[\sin \omega t - (\sin 3\omega t)/3^2 + (\sin 5\omega t)/5^2 - \ldots] = \begin{cases} 4t/T & -T/4 \leqslant t \leqslant T/4 \\ 2 - 4t/T & T/4 \leqslant t \leqslant 3T/4. \end{cases}$$

Sawtooth Waveforms

$$(2/\pi)[\sin \omega t - (\sin 2\omega t)/2 + (\sin 3\omega t)/3 - \ldots] = 2t/T \qquad -T/2 < t < T/2$$

$$(-2/\pi)[\sin \omega t + (\sin 2\omega t)/2 + (\sin 3\omega t)/3 + \ldots] = -1 + 2t/T \qquad 0 < t < T.$$

Half-wave Rectified Cosine and Sine Waveforms

$$\frac{1}{\pi} + \frac{\cos \omega t}{2} + \frac{2}{\pi} \sum_{n=1}^{\infty} (-1)^{n+1} \frac{\cos 2n\omega t}{(4n^2 - 1)} = \begin{cases} \cos \omega t & -T/4 \leqslant t \leqslant T/4 \\ 0 & T/4 \leqslant t \leqslant 3T/4 \end{cases}$$

$$\frac{1}{\pi} + \frac{\sin \omega t}{2} - \frac{2}{\pi} \sum_{n=1}^{\infty} \frac{\cos 2n\omega t}{4n^2 - 1} = \begin{cases} \sin \omega t & 0 \leqslant t \leqslant T/2 \\ 0 & T/2 \leqslant t \leqslant T. \end{cases}$$

Rectified Cosine and Sine Waveforms

$$\frac{2}{\pi} - \sum_{n=1}^{\infty} (-1)^n \frac{4 \cos 2n\omega t}{\pi(n^2 - 1)} = \cos \omega t \qquad 0 < t < T$$

$$\frac{2}{\pi} - \sum_{n=1}^{\infty} \frac{4 \cos 2n\omega t}{\pi(n^2 - 1)} = \sin \omega t \qquad 0 < t < T.$$

Train of Rectangular Pulses

$$\frac{2a}{T} + \sum_{n=1}^{\infty} \frac{2 \sin n\omega a \cos n\omega t}{\pi n} = \begin{cases} 0 & -T/2 < t < -a \\ 1 & -a < t < +a \\ 0 & a < t < T/2 \end{cases}$$

Appendix 2: Some Fourier Transforms

In the following, $y(t)$ and $Y(f)$ are related by equations (2.15) and (2.16). At a discontinuity, each function is assigned a value midway between the limiting values each side of the discontinuity. Throughout, a represents any positive number, whilst b represents any real number (positive, negative, or zero).

Rectangular Pulse

$$y(t) = \begin{cases} 1 & (b-a) < t < (b+a) \\ 0 & \text{otherwise} \end{cases} \qquad Y(f) = \frac{\sin(2\pi a f)}{\pi f} \exp(-2\pi i b f).$$

Pair of Rectangular Pulses

$$y(t) = \begin{cases} 1 & (b-a) < |t| < (b+a) \\ 0 & \text{otherwise} \end{cases} \qquad Y(f) = \frac{2\cos(2\pi b f)\sin(2\pi a f)}{\pi f}.$$

$[b > a]$

Triangular Pulse

$$y(t) = \begin{cases} 1 - |t|/a & -a < t < a \\ 0 & \text{otherwise} \end{cases} \qquad Y(f) = a\left(\frac{\sin \pi a f}{\pi a f}\right)^2.$$

Double-sided Exponential

$$y(t) = \exp(-a|t|) \qquad -\infty < t < \infty \qquad Y(f) = \frac{2a}{a^2 + 4\pi^2 f^2}.$$

Single-sided Exponential

$$y(t) = \begin{cases} \exp(-at) & t > 0 \\ 0 & t < 0 \end{cases} \qquad Y(f) = \frac{1}{a + 2\pi i f}.$$

Gaussian

$$y(t) = \exp(-a^2 t^2) \qquad -\infty < t < \infty \qquad Y(f) = \frac{\sqrt{\pi}}{a} \exp(-\pi^2 f^2/a^2).$$

Cosine Pulse

$$y(t) = \begin{cases} \cos bt & -a < t < a \\ 0 & |t| > a \end{cases}$$

$$Y(f) = \frac{\sin 2\pi a(f-b)}{2\pi(f-b)} + \frac{\sin 2\pi a(f+b)}{2\pi(f+b)}$$

Shifting, Modulation and Scaling

Suppose $y(t)$ and $g(t)$ have transforms $Y(f)$ and $G(f)$. It will follow that:

(i) if $g(t) = y(t-b)$ then $G(f) = \exp(-2\pi i b f) Y(f)$
(ii) if $g(t) = \exp(2\pi i b t) y(t)$ then $G(f) = Y(f-b)$
(iii) if $g(t) = a y(at)$ then $G(f) = Y(f/a)$.

Index

AC theory, 24
Amplitude, 1, 3, 4
Amplitude spectrum, 7
Angular frequency, 1
Angular wavenumber, 9
Average value, 11

Band-pass filter, 26
Bandwidth theorem, 10, 36–8
BASIC program, 42, 43
de Broglie relation, 10, 37

Cauchy–Riemann limit, 48
Complex Fourier series, 17–18
Complex notation, 16–18
Computer program, 42, 43
Conduction of heat, 34–6
Continuous amplitude spectrum, 7
Cooley, J W, 43
Cosine amplitude, 3, 4
Cosine amplitude spectrum, 7, 8
Cosine coefficient, 11
Cosinusoidal oscillation, 1

DFT, 44
Diffraction, 9, 26–31
  Fraunhofer, 31
  grating, 10
Discrete amplitude spectrum, 8
Discrete Fourier transform, 44
Dirichlet conditions, 47, 48

Electrical circuits, 24–6
Electron diffraction, 37
Energy dissipation, 22
Even functions, 15, 20

Fast Fourier transform, 42–4
FFT, 43

Filters, 25
  low-pass, 25
  band-pass, 26
Fourier, J B, 4
Fourier analysis, 1, 4, 19
Fourier formulae, 11, 16, 18, 19
  proof of, 46–9
Fourier, integral, 11, 19
Fourier inversion theorem, 19, 47–8
Fourier series, 11, 17
  coefficients, 11, 17, 18
  computation of, 39, 40, 44
  complex, 17, 18
  half-range, 15, 16, 32, 34
  list of, 51, 52
  theorem, 11, 46–8
Fourier synthesis, 19, 45
Fourier transforms, 19
  computation of, 41–2, 45
  fast, 42–4
  list of, 53–4
  theorem, 19, 47–8
Fraunhofer diffraction, 31
Frequency, 1
  angular, 1
  fundamental, 4
  negative, 17
  spatial, 9
Fundamental frequency, 4

Gaussian function, 37, 38, 54

Harmonic
  analysis, 3, 4
  frequencies, 4
  oscillation, 1, 3, 16
  synthesis, 3, 4
Half-range Fourier series, 15, 16, 32, 34
Half-wave rectified oscillation, 14, 52

Heat conduction, 34–6
Huygens' principle, 27, 29, 31

Impedance, 24

Light, 9
Low-pass filter, 25

Mean square value, 22
Mean value, 11
Mode, normal, 32
Momentum, 10
  probability, 37
Monochromatic light, 9

Negative frequency, 17
Normal mode, 32
Numerical integration, 39

Odd functions, 15, 20
Oscillations
  cosinusoidal, 1
  harmonic, 1, 3, 16
  on a string, 31–4
  periodic, 4, 8
  sinusoidal, 1

Parseval's formulae, 22–3, 48–9
Partial sum, 5
Pathological behaviour, 11, 22, 23
Pendulum, 3
Period, 1
Periodic oscillation, 4, 8
Phase, 1
  ambiguity, 3
Planck's constant, 10, 37
Power dissipation, 22
Pressure, 9
Program, computer, 42, 43
Pulse, 9
Pure tone, 9

Rectangular pulse, 20, 53
Rectified waveform, 14, 52
Riemann–Cauchy limit, 48

Sampling, 39, 41
Sampling error, 42
Sawtooth oscillation, 6, 12, 18, 25, 51
Schrödinger wave function, 10, 37–8
Signal, 9
Sine amplitude, 3, 4
  spectrum, 7
Sine coefficient, 11
Sinusoidal oscillation, 1
Sound wave, 9
Spatial frequency, 9
Spectrum,
  continuous, 7
  cosine and sine, 7, 8
  optical, 9
Square pulse, 20, 53
Square waveform, 12, 25, 51
String, oscillation of, 31–4
Synthesis,
  Fourier, 19, 45
  harmonic, 3, 4, 9

Thermal conductivity, 34
Transfer function, 25
Transient, 8
Transmission function, 27
Transmittance, 9
Triangular pulse, 53
Triangular waveform, 51
Truncation, 41
  error, 42
Tukey, J W, 43

Uncertainty principle, 10, 36–8

Wavelength, 9
Wave mechanics, 10
Wavenumber, 9
  angular, 9
Wave packet, 37

Young's experiment, 27, 29